Techonomics

C. A. Chuba

VITAE PRESS
EMBARK ON YOUR QUEST

ISBN: 978-1-967470-12-9

Published by Vitae Press LLC

Cover design by C. A. Chuba

Vitae Press

Vitae Press is a publishing house dedicated to exploring the enduring legacy of Western life, thought, and culture as shaped by the Roman tradition. The name *Vitae*, derived from the Latin word for "life," reflects our mission to publish works that engage with the values, ideas, and heritage of the Western world. From fiction to non-fiction to reprints of timeless works from Christendom and Western Civilization, each book seeks to deepen the understanding of the moral, spiritual, and intellectual foundations that have guided generations and continue to inspire today.

TABLE OF CONTENTS

INTRODUCTION

This book is neither a doomsday prediction of artificial intelligence catastrophe nor a celebration of Silicon Valley optimism. It does not claim that technology will save us, nor does it assume that technological progress automatically leads to human flourishing. For many years, I believed, as many do, that technology was simply one sector of the modern economy, comparable to healthcare, insurance, banking, or energy. Like millions of Americans, I thought of "tech" as something happening far away in California, among programmers and venture capitalists whose work felt disconnected from the daily concerns of ordinary life. The last decade proved that assumption deeply mistaken.

What we casually call technology has evolved into the digital infrastructure of modern civilization itself. It is no longer one industry among many. It has become an ecosystem that shapes finance, communication, education, healthcare, entertainment, manufacturing, governance, and even personal identity. Technology simplifies tasks and increases efficiency, but it also displaces workers, restructures institutions, and redefines the foundations of economic and social life. This transformation, and the economic order created by it, is what this book examines under the concept of *Techonomics*.

Modern technologies offer remarkable advantages. Precision navigation, instantaneous knowledge retrieval, and personalized healthcare, finance, and nutrition supported by advanced artificial intelligence promise unprecedented efficiency. Yet these same systems enable surveillance at a scale unknown to previous civilizations. Online behavior, eye movement, location, household activity, and biometric signals are continuously tracked. Homes, vehicles, and workplaces generate data by default. These systems collect personal information with a level of granularity that no prior generation of Western civilization would have accepted as normal. Capability and vulnerability now advance together.

This shift unfolds while much of the world is still living amid the unresolved consequences of the Industrial Revolution. Polluted rivers, contaminated soil, and microplastics now found in human organs are not relics of a distant past. They are present realities. Globalized production provides low-cost goods, but it has also hollowed out local manufacturing, weakened community institutions, and left entire regions marked by abandoned factories and economic stagnation. Unemployment, addiction, demographic decline, and social fragmentation persist outside major metropolitan centers. The digital transformation does not begin from a position of stability. It accelerates atop an already strained cultural and economic foundation.

Over the past decade, this transformation has intensified at a speed without precedent in modern economic history. Technologies once confined to academic laboratories or niche online communities have become engines of global influence. Bitcoin began as an obscure experiment and became a challenge to traditional assumptions about money, authority, and financial sovereignty. Artificial intelligence advanced with similar acceleration. It now drafts reports, generates code, edits media, designs products, and curates' information consumed by billions. These tools no longer exist at the margins of society. They have become embedded in the operating systems of daily life.

Most people encounter these systems without recognizing their influence. They scroll through feeds, accept recommendations, and consume content optimized to capture attention. Behind every notification or suggestion, artificial intelligence sorts, filters, and elevates ideas according to incentives that reward engagement rather than truth. Automated systems amplify tension, shape opinion, and quietly guide behavior. What appears as personal choice is often the output of invisible optimization.

The digital environment extends far beyond familiar platforms. Machine learning models influence journalism, advertising, entertainment, and financial decisions. Studios rely on automated systems to enhance or generate content that blurs the line between authentic and synthetic. Notifications and prompts follow users continuously. The boundary between the physical world and the digital layer that observes, predicts, and interprets it has become increasingly thin. Daily life unfolds under constant data extraction.

At the same time, traditional currencies lose purchasing power while digital assets governed by code and networks gain legitimacy. Individuals who adapt early often experience new forms of financial mobility. Those who do not remain bound to institutions shaped by industrial era assumptions. Wealth increasingly concentrates around firms that control data pipelines, cloud infrastructure, and artificial intelligence platforms. Economic power shifts from ownership of physical assets toward management of digital systems.

Data has become the central resource of the modern economy. It directs advertising, informs credit decisions, influences politics, and guides corporate strategy. Algorithms determine which companies rise, which ideas circulate, and which opportunities appear before individuals. Digital assets introduce programmable money and decentralized ownership structures. Artificial intelligence accelerates these changes beyond human cognitive speed. Financial technology embeds economic power directly into the devices people rely on each day.

Traditional institutions struggle to keep pace. Regulation is often reactive, fragmented, and incomplete. In this environment, digital literacy becomes a form of economic capital. Those who can interpret technological signals and adapt their strategies are rewarded. Those who rely on older assumptions about work, education, and stability face increasing strain as automation reshapes labor markets and digital platforms accumulate disproportionate influence.

History offers perspective, though not comfort. The Industrial Revolution displaced workers, reordered classes, and reshaped political structures. The digital revolution is broader and faster. It transforms not only labor and industry, but authority, identity, and value itself. Lifelong learning is no longer optional. Curiosity and adaptability replace fixed expertise. Those who understand the landscape may navigate it. Those who ignore it will still be shaped by it.

What must now be stated plainly is this: movements do not remain free indefinitely. Either a movement conquers the system, or the system absorbs and neutralizes the movement. History offers no third option.

The technological economy described in this book did not emerge suddenly. It is the mature form of a process set in motion at the beginning of the twentieth century. The Industrial Revolution did not disappear. It entered a cocoon. Like the caterpillar, it consumed raw materials, reshaped landscapes, erected factories, mobilized populations, and reorganized nations. That process culminated in the World Wars, which were not historical accidents, but consequences of industrial power driven to its limit.

What followed was transformation rather than resolution. Industrial infrastructure remained intact, but its purpose shifted. Steel, oil, electricity, logistics, and mass production became the foundation for a new form of power. The butterfly emerged. It was still the caterpillar, but perfected, faster, and more precise. The modern tech economy depends entirely on industrial raw materials and physical infrastructure, yet it serves a different end. Power is no longer exercised primarily through bombs and battalions, but through digital control, informational leverage, and economic coordination at scale.

The financial collapse of 2008 and 2009 marked a decisive inflection point. Public trust fractured as institutions that preached discipline were rescued without consequence. Firms that had leveraged themselves aggressively in pursuit of extraordinary returns were shielded from failure with public funds. Gains remained private. Losses were transferred to taxpayers. Markets were prevented from correcting themselves, insulating those who took excessive risks while transferring the cost of failure to the public.

It was in this atmosphere that a series of technical manifestos appeared, the most consequential being the Bitcoin white paper. These documents were written in the language of cryptography and distributed systems, but their motivation was not merely technical. They proposed an alternative to discretionary financial power. A system in which trust was replaced by verification, issuance by rule, and participation by mathematics rather than permission. The white papers outlined an alternative architecture, one designed to operate outside institutions that had proven unwilling to discipline themselves. They assumed that institutions distorted by moral hazard could not be reformed, only bypassed.

This intellectual rebellion was mirrored in the physical world. The *Occupy Wall Street* movement gave public voice to the same intuition. That financial systems no longer rewarded productive risk, that leverage had been decoupled from consequence, and that political representation had become subordinate to capital preservation. The slogans were uneven and the organization fragile, but the diagnosis resonated. A generation sensed that the system was no longer neutral, and that access itself had become conditional.

For a brief period, resistance appeared plausible. Cryptographic money circulated outside banks. Open-source networks grew without corporate ownership. Peer-to-peer systems promised coordination without hierarchy. The counter-movement was fragmented, but it was real. Then the pattern reasserted itself.

Large institutions adapted faster than individuals. Corporations integrated digital assets, artificial intelligence, and financial technology into existing power structures. Blockchain became infrastructure rather than rebellion. Artificial intelligence displaced labor while concentrating ownership. Fintech reduced friction without redistributing authority. Political movements promised disruption but delivered continuity, translating protest into policy language that left underlying systems intact. What began as an effort to escape the system is now being absorbed into it.

The distance between ordinary participants and the mechanisms governing their lives has widened again, not primarily through exclusion, but through complexity. Systems grow more powerful as they become less visible. Decision-making migrates into code, policy frameworks, and administrative processes few can inspect or challenge. Power remains, but it is exercised indirectly and without a clear face.

This book is written because history shows where this pattern leads. When ownership is removed from common understanding, when participation requires comprehension beyond the reach of ordinary citizens, society drifts toward dependency. Not tyranny in its classical form, but administration. Not chains, but permissions.

The purpose of this work is not to alarm, nor to console. It is to explain. To trace the logic of *Techonomics* from its industrial origins to its digital maturity. To show how movements arise, how systems respond, and how power adapts. Those who understand this process may yet retain agency. Those who do not will live inside it without knowing how it came to be. History does not ask permission. It only asks whether we are prepared.

C. A. Chuba
Pittsburgh, Pennsylvania
December, A.D. 2025

TECHONOMICS

The Crypto, AI, DeFi, and Fintech Revolution Explained

Chapter I

A Century of Digital

The Birth of Computing (1930s-1970s)

To understand *digital assets, Artificial Intelligence (AI), decentralized finance (DeFi)*, and modern *fintech*, we must begin at the moment when information first became something machines could process. The digital age did not begin with smartphones or the internet. It began with the idea that thought itself could be expressed as rules, reduced to symbols, and carried out by a device. That single shift blurred the boundary between human reasoning and mechanical procedure. It opened the door to extraordinary possibilities while hinting at a future in which civilization might rely as much on machines as on people.

The story of modern *computing* begins with ideas that sounded theoretical long before they became hardware. In 1936, the English mathematician Alan Turing (1912–1954) introduced the concept of a *Turing machine*, a hypothetical device that follows step-by-step rules to manipulate symbols. His work proposed that information could be treated mathematically and processed logically according to instructions, independent of any particular physical machine. These ideas became a blueprint for digital computing.

Turing imagined a device that operated on a long strip of tape divided into squares, with each square containing a symbol such as a number or a letter. A reading head would move along the tape, scanning one square at a time, while a list of instructions directed the machine to read a symbol, erase it, replace it, or move to another square. By repeating this process step by step, the device could carry out a complete calculation. The idea was almost childlike in its simplicity, yet it demonstrated that a machine could perform any logical operation if the rules were precise enough.

This concept mirrors the structure of modern computing. Computers receive information, examine it piece by piece, and modify it according to rules written in code. Even the most advanced programming languages follow the same principle. When someone opens an app, runs software, or searches the internet, a computer is carrying out countless tiny actions modeled on the logical framework Turing outlined in the 1930s. His thought experiment showed that the essence of computing lies less in wires and hardware than in logic itself.

During the Second World War (1939–1945), these concepts moved from theory into practice. At Bletchley Park in Buckinghamshire, England, Turing and his colleagues used early electromechanical systems to attack encrypted German communications produced by the *Enigma machine*. Enigma was a portable cipher device used by the German military that scrambled messages through rotating internal wheels and electrical circuits, creating an enormous number of possible letter combinations.

Each keystroke on Enigma substituted one letter for another, but the substitution changed continuously, producing codes that were practically impossible to break by hand. The machines at Bletchley Park, which filled entire rooms and required teams of operators, proved that once information could be expressed mathematically, mechanical systems could search possibilities and execute operations far faster than human codebreakers. Wartime *cryptography* revealed that computing was not merely an intellectual exercise. It had strategic value.

After the war, engineers began constructing programmable computers, meaning machines that could be instructed to perform different tasks by changing their internal instructions. One of the first and most influential was *Electronic Numerical Integrator and Computer (ENIAC)*, completed in 1945 in the United States at the University of Pennsylvania. It was built for the U.S. Army to calculate artillery firing tables by solving complex equations involving trajectory, wind, and distance. Calculations that once took teams of human mathematicians days or weeks could now be performed electronically in minutes or hours.

Although programming ENIAC required operators to physically plug and unplug cables, change switches, and rewrite its configuration, it represented a decisive shift. Instead of building a different machine for each new task, engineers discovered that a single system could be reconfigured to solve entirely new problems. ENIAC soon moved beyond artillery tables into scientific modeling, nuclear research calculations, and mathematical simulation. This flexibility marked the transition from specialized calculators to general-purpose computers, the direction that led to modern software, personal computers, and today's digital platforms.

During the 1950s and 1960s, the next major leap followed from the invention of the *transistor* in 1947 by researchers at Bell Labs. A transistor is a tiny component that can act as a switch or amplifier for electrical signals. It controls the flow of electricity in a circuit, allowing information to be represented as patterns of on and off states. Before transistors, computers relied on vacuum tubes, which were large, fragile, heat-producing, and prone to failure. Transistors replaced them with something smaller, more durable, and far more efficient.

That breakthrough was followed by the *integrated circuit*, developed in 1958 by Jack Kilby (born 1923) and independently by Robert Noyce (born 1927). An integrated circuit places many transistors on a single piece of *silicon*, forming a compact electronic system rather than a collection of separate parts. Logical operations that once required large assemblies could now occur inside a tiny chip. As engineers packed more transistors into smaller spaces, computers that once filled rooms were reduced to cabinets, then desktops, and eventually devices small enough to fit on a table or inside a pocket. These innovations made modern computing possible and set the stage for the personal computer revolution of the late twentieth century.

As computing matured, the center of technological development increasingly shifted toward California, particularly the region south of San Francisco Bay. During the 1950s and 1960s, *semiconductor* firms clustered there to manufacture transistors and integrated circuits, both dependent on silicon as the base material for electronic chips. Engineers, many connected to Stanford University, founded companies focused on semiconductor research and chip production. As startups, laboratories, and manufacturing facilities concentrated in the same area, the region took on a distinct identity. Journalists and local promoters began calling it *Silicon Valley*, a reference to both the physical material inside computer chips and the growing community of innovators shaping the future of electronics.

By the 1970s, computers had become faster, cheaper, and accessible to institutions outside government and the military. Universities, research centers, and early technology companies began using digital systems for science, business, and communication. Computing moved from theoretical mathematics and wartime urgency into a practical tool that would soon enter offices, schools, and eventually private homes.

The Creation of the Internet (1960s-1990s)

The internet, understood today as a global network of interconnected computers that exchange data through shared rules, began as a United States government research effort in the 1960s. The U.S. Department of Defense funded *Advanced Research Projects Agency Network (ARPANET)*, which linked university and military research computers so they could exchange information over long distances. One of its original goals was network resilience, meaning that if one part of the system failed, information could still travel through remaining connections. This introduced a new idea in computing: machines cooperating across distance rather than functioning as isolated devices.

ARPANET was administered by *Advanced Research Projects Agency (ARPA)*, a branch of the Department of Defense created in 1958 in response to the Soviet launch of Sputnik. Its mission was to prevent technological surprise. Instead of developing technology solely inside military institutions, ARPA funded researchers at universities and private laboratories, allowing experimental ideas to be explored with government support. Over time, the agency moved deeper into networking, interactive computing, and early artificial intelligence research.

In 1972, ARPA was renamed *Defense Advanced Research Projects Agency (DARPA)*, reflecting its expanded role. Many foundational technologies of the modern internet emerged from the research ecosystem supported by ARPA and later DARPA. One of the earliest practical uses of ARPANET was electronic mail (email), introduced in the early 1970s. Researchers in different places could send messages to one another in seconds instead of relying on physical mail or internal memos. This single application showed that networks were not only tools for transferring files. They were becoming a medium of daily communication.

To make different networks compatible, engineers developed *Transmission Control Protocol/Internet Protocol (TCP/IP)*, a family of communication rules that break information into packets, send them across multiple networks, and reassemble them on arrival. A practical example is loading a web page, where text and images are split into packets, routed across many systems, and rebuilt in a browser. In 1983, ARPANET adopted TCP/IP, allowing separate networks to interconnect. This step transformed isolated systems into a true network of networks and laid the technical foundation of today's internet.

In 1989, the English computer scientist Tim Berners-Lee (born 1955) proposed a new way to organize and connect information stored on different computers. At the time, he was working at *CERN*, the European Organization for Nuclear Research, a physics laboratory on the border of Switzerland and France. His idea became the foundation for what we now call the *World Wide Web (WWW)*.

CERN employed thousands of scientists from around the world, and the growing volume of digital documents made collaboration difficult. Berners-Lee created a method of linking documents through hypertext, where selecting a link could take a user to another document stored elsewhere on the network. To make this workable for ordinary users, he built the first web browser and web server, allowing documents to be retrieved and viewed through an interface rather than through specialized terminal commands.

A useful distinction is that the Web runs on top of the Internet rather than being identical to it. The Internet is the global network of computers and physical infrastructure that moves data. The Web is a layer of software conventions that makes that data easy to access and navigate. Without the Internet, the Web could not operate. Without the Web, the Internet would have remained largely confined to specialists. Once Berners-Lee demonstrated that users could open pages, follow links, and move through information visually, the internet shifted from a research network into a public platform.

By the early 1990s, browsers such as Mosaic and Netscape made the Web easier to navigate, and early search engines, including Yahoo and later Google, helped users locate information. Businesses, news organizations, and individuals began building websites. The internet moved from an academic and military project into a commercial and social infrastructure that would support trade, media, and everyday communication.

Web 2.0 and Social Media (2000s)

In the early 2000s, the internet entered a phase often called *Web 2.0*, a period in which websites became interactive and user driven rather than static and publisher controlled. During the earlier *Web 1.0* period of the 1990s, most sites consisted of fixed pages created by small groups of authors and offered limited interaction. Web 2.0 expanded participation at scale, allowing ordinary users to post, comment, review, rate, and collaborate directly online. The shift was visible in ordinary life. Instead of merely reading webpages, people posted updates, edited shared documents, uploaded videos, and built audiences.

Several developments made Web 2.0 possible. *Broadband internet* replaced older *dial-up* connections for many households. Broadband transmitted data continuously rather than through a phone-line modulator, allowing faster speeds and smoother handling of images and video. It also made constant connection normal. At the same time, techniques such as *Asynchronous JavaScript and XML (AJAX)* allowed a web page to update part of its content without reloading the entire screen. Users could interact with menus, comment boxes, maps, and galleries with far less delay.

These improvements made websites behave more like software applications. Publishing tools such as Blogger and WordPress allowed people to write and format content without knowing *HyperText Markup Language (HTML)*. A person no longer needed to write code by hand to format titles, insert pictures, or create links. Platforms such as Flickr supported large-scale photo sharing, making visual content part of everyday online life. The web began shifting from a digital library people visited into a dynamic environment shaped by users themselves.

During this period, Google came to dominate search, the process of finding relevant pages across the web. It relied heavily on *algorithms*, step-by-step procedures that process information to produce results, such as ranking pages by relevance. Search advertising, the sale of placement in or alongside results, provided a business model that funded free services while turning user attention into revenue.

The most visible expression of Web 2.0 was social media, online platforms where users maintain profiles, connect with others, and share content. Facebook enabled personal networks. YouTube made video publishing available to anyone. Twitter distributed short messages in real time. Platforms such as MySpace experimented with music sharing and personal pages. In each case, users supplied the content while the platform controlled discovery and distribution.

Central to these platforms was the rise of recommendation algorithms, systems that select which posts, videos, or links a user is most likely to engage with. A video platform tracks what a user watches and suggests similar material to keep the user on the site. These systems shaped what people saw, how information spread, and which voices were amplified. They laid the groundwork for the *attention economy*, a market in which human attention is treated as a scarce resource and competed for.

By the late 2000s, *digital identity* had become a central feature of social life. Posts, profiles, and follower counts became signals of status. At the same time, each click and interaction generated data that could be analyzed and monetized. Web 2.0 did not merely improve website design. It altered how people related to one another, how businesses reached customers, and how influence traveled through society.

Mobile, Cloud, and Platform Economies (2010s)

The 2010s were defined by the spread of mobile computing, access to digital services through portable devices such as smartphones and tablets. When Apple released the iPhone and competitors followed, a powerful computer moved into the average person's pocket. The app store model, digital marketplaces where users download software directly to their devices, allowed developers to reach global audiences with relatively low upfront cost. A ride-sharing app that connects drivers and passengers in real time became a familiar example.

At the same time, *cloud computing* became the backbone of digital services. Cloud computing delivers storage and processing power over the internet from remote data centers. Companies such as Amazon Web Services, Microsoft Azure, and Google Cloud built large clusters of servers that other organizations could rent rather than own. A startup could launch a new service by paying for capacity as needed instead of buying hardware. Computing became a utility available on demand.

Cloud infrastructure also enabled *big data*, collections of information so large and complex that traditional tools cannot process them efficiently. As firms stored detailed records of online behavior, transactions, and sensor readings, researchers expanded methods for machines to recognize patterns directly from data. This approach drew on early artificial intelligence research from the 1950s, when figures such as Arthur Samuel (born 1901) experimented with programs that improved through experience rather than relying only on fixed rules.

Over time, these techniques were refined into *machine learning (ML)*, a branch of AI in which models learn patterns from data and use them to make predictions without being explicitly programmed for every scenario.

An everyday example is an online retailer predicting which products a customer might want based on prior behavior. Together, mobile devices, cloud computing, and machine learning allowed firms to personalize services, optimize logistics, and automate decisions at a scale that traditional programming could not achieve.

The decade also saw the rise of the *platform economy*, a business model in which a company provides a digital marketplace connecting buyers and sellers, or providers and users. Amazon connected merchants and customers. Uber linked drivers with passengers. Airbnb connected hosts with guests. Netflix delivered digital content to subscribers. These companies typically own software and data rather than physical assets. A ride-sharing firm does not need to own cars to influence transportation flows in a city.

Platforms benefited from *network effects*, a dynamic in which a service becomes more valuable as more people use it. The more buyers who visited a marketplace, the more attractive it became for sellers, which drew still more buyers. Combined with mobile access and cloud infrastructure, network effects allowed a small number of firms to scale rapidly and dominate entire sectors.

By the end of the decade, daily life for many people involved constant interaction with mobile apps, cloud-based services, and platforms that mediated shopping, transport, communication, and entertainment. The foundations of a new economic order were in place, one centered on data, connectivity, and software rather than on factories and heavy industry.

The Breaking Point (2015-2025)

Between 2015 and 2025, several technological forces converged in ways that shifted economic power. Digital platforms, AI, cloud infrastructure, mobile networks, and digital assets became tightly integrated into daily life. What had once appeared as a distinct sector increasingly functioned as the organizing layer of the entire economy. In this period, value creation moved away from industrial machinery toward data systems and software networks.

The idea of a decentralized digital environment gained formal expression when Gavin Wood coined the term *Web 3.0* in 2014, shortly after co-founding the blockchain platform Ethereum (ETH). The concept did not enter mainstream discussion until around 2021, during the COVID-19 pandemic, when large parts of the global population were pushed into remote work, online communication, and digital financial systems. The pandemic did not create this shift, but it exposed how dependent modern economies had already become on digital infrastructure. It marked a visible breaking point in the transition from industrial systems to *Techonomics.*

As Web 2.0 concepts migrated into the emerging Web 3.0 environment, familiar terms were replaced with vocabulary built for decentralized systems. In Web 2.0, most activity runs on servers owned by companies. In Web 3.0, core functions are spread across distributed networks operated by many independent participants. As a result, language shifted alongside architecture. Servers became *blockchains*. Traditional databases became *distributed ledgers*. Paying for server resources became paying *gas fees*. The Web 2.0 server administrator was replaced by *miners*, *validators*, or *stakers*. Individual machines running the network became *nodes*.

Applications also shifted. Web 2.0 services typically rely on centralized platforms such as Google Cloud or Amazon Web Services. Web 3.0 applications rely on *smart contracts*, programs that execute on blockchain networks according to predetermined rules. Publishing content became *minting* assets onto a blockchain. Passwords became *private keys*. Simplified recovery systems became *seed phrases*. A platform login became a *digital wallet*, controlled by *cryptographic keys* rather than a corporate account. Media files that once existed as ordinary digital objects became *non-fungible tokens (NFTs)*, meaning ownership could be verified on a ledger rather than being defined solely by a platform's database.

Financial activity followed the same pattern. Bank accounts, brokerage accounts, and credit card systems translated into digital wallets, exchanges, and *tokenized assets*. Platform-specific assets became *tokens*. The process of digitizing real-world claims became *tokenization*, meaning that assets such as art, land titles, or financial holdings can be represented as tokens that trade globally without intermediaries. Scaling through larger centralized server capacity was increasingly discussed through concepts such as *sharding* or *layer-2 solutions*, approaches that distribute computation across multiple systems rather than processing everything on a single chain.

Taken together, these changes show that Web 3.0 was not merely a rebranding of earlier internet technology. It represented the construction of a new technical layer. Functions that Web 2.0 handled through centralized servers, platform accounts, and corporate databases began shifting into decentralized networks, cryptographic identity, and programmable assets. Once this infrastructure existed, it provided a foundation for new forms of digital value to emerge, and for the movement of value, computation, and ownership to occur with reduced dependence on any single institution.

At the same time, AI entered a new phase. *Generative AI* models, systems capable of producing text, images, audio, and code from patterns learned in data, became widely available to the public rather than remaining confined to universities and research laboratories. For decades, AI progress appeared distant from ordinary life. Beginning around 2022, tools capable of writing essays, drafting messages, answering questions, and producing images reached everyday users through simple interfaces. Programs such as ChatGPT, DALL-E, and Midjourney demonstrated that advanced language and image generation could be accessed without specialized training, and what once appeared experimental began integrating into daily tasks and professional environments.

These systems depended on more than clever software. Large cloud providers supplied the computing power required to train and run them. Global data centers processed the enormous datasets on which they relied. Conversational models and image generators showed that capable AI could be deployed at global scale, interacting with millions of users, responding instantly, and improving through expanded training. As these tools spread, AI began supporting or replacing tasks in marketing, customer service, law, finance, and healthcare. Occupations that had once been primarily human became hybrid environments in which machines performed cognitive labor alongside people.

Everyday devices such as smartphones, laptops, and home assistants connected billions of people to these systems. Individuals could participate in digital asset markets, use AI-supported tools, and interact with services that recorded and analyzed their activity. Corporations and governments expanded data infrastructures to accommodate these applications. The boundary between the physical economy and the digital layer became less distinct as ownership records, contracts, medical information, personal communications, and professional identity moved into digital form and became subject to algorithmic processing.

The result is an environment in which computing power, global connectivity, data, digital assets, and AI-driven systems operate together. What began as separate revolutions in computing, networking, finance, and machine learning has matured into a single tech economy, an order organized primarily around digital infrastructure and software platforms rather than mechanical production. In such a system, advantage flows to those who understand the mechanisms, control the infrastructure, or can interpret the data these systems generate.

The chapters that follow examine this transformation in greater detail, including digital assets, decentralized finance (DeFi), AI, and the economic and social consequences of living in a world shaped primarily by digital systems rather than by industrial machinery.

Chapter II

Techonomics Rising

The New Economic Paradigm

For most of modern history, classical economics described an economy built on land, labor, and capital, the physical resources, human effort, and industrial machinery that produced tangible goods. A factory, a farm, and a shipping fleet were the unmistakable symbols of wealth. Growth was constrained by scarcity. A farm could harvest only what the soil could sustain. A factory could produce only what its machinery and workers could handle. Every additional unit of output required more people, more materials, or more machines. Value remained tethered to geography, distance, and physical effort.

Today the engine of economic power is different. In the digital economy, the central levers are data, algorithms, and networks. *Data* is recorded behavior, what people search for, buy, watch, type, scan, or track.

Algorithms are step-by-step computational instructions that determine what to recommend, approve, deny, or display. *Networks* are the interconnected systems that move this information across billions of devices. Every click, payment, GPS signal, or biometric scan becomes raw material that can be harvested, analyzed, and transformed into prediction, automation, or influence. Wealth creation has shifted from extracting natural resources to extracting behavioral information.

Governments around the world have acknowledged this shift by redirecting policy and national strategy toward data infrastructure, semiconductor capacity, cloud computing, artificial intelligence, and digital financial systems. In the United States, this became especially visible during the 2020s through a series of major laws and federal initiatives.

One recent example is the *Guiding and Establishing National Innovation for United States Stablecoins Act*, known as the *GENIUS Act*, which President Donald J. Trump signed into law on July 18, 2025. The Act establishes a comprehensive federal framework for payment *stablecoins*. It sets requirements for reserve backing, redemption rights, supervision, and anti-money-laundering controls. It allows both federal and state regulatory pathways, brings stablecoin issuers under safety and soundness standards similar to banking, and integrates digital dollar payment infrastructure into mainstream financial oversight. The purpose of the law is to create a trusted environment for dollar-based digital assets while expanding the United

States role in the global crypto economy.

The GENIUS Act builds on earlier efforts. One of the most significant was the *CHIPS and Science Act of 2022*, signed on August 9, 2022. CHIPS refers to *Creating Helpful Incentives to Produce Semiconductors.* The law authorized roughly 280 billion dollars in funding and incentives for domestic semiconductor manufacturing, advanced materials research, scientific innovation, workforce development, and national security-related technologies. It directed the Department of Commerce, the National Science Foundation, and other agencies to expand research hubs, strengthen supply chains, and promote long-term leadership in critical technologies such as artificial intelligence, quantum computing, and high-performance computing. It also aimed to reduce American dependence on foreign semiconductor production, which had become a strategic vulnerability during the global supply chain shocks of the early 2020s.

Federal agencies also pursued programs designed to broaden access to AI research capacity. The *National AI Research Resource (NAIRR)* emerged in early 2020s policy planning as a shared national infrastructure intended to provide compute power, curated datasets, and testing environments for researchers, universities, and smaller firms. Its goal is to ensure that innovation is not limited to a small number of technology companies that control most advanced computing resources. It represents an effort to democratize artificial intelligence research while still guiding it through coordinated national priorities.

In addition to legislation, the federal government used executive authority to steer AI development as a matter of national strategy. *Executive Order 14110*, issued on October 30, 2023 by President Joe Biden, established a policy framework for the safe, secure, and trustworthy development and deployment of artificial intelligence. The order instructed agencies to develop standards for AI safety, address algorithmic discrimination, strengthen privacy protections, prepare the workforce for AI-driven economic shifts, and expand research into responsible innovation. Although later administrations may revise or redirect portions of the order, it demonstrated the degree to which artificial intelligence had become a top-level federal concern touching national security, healthcare, finance, transportation, and civil rights.

Together, these policies illustrate a fundamental change in how governments operate in the digital era. They no longer merely observe technological progress. They cultivate digital infrastructure, fund scientific research, create regulatory frameworks for new financial systems, and position AI and cryptographic technology as instruments of national competitiveness. Public policy, once focused on agriculture, manufacturing, or trade, now treats software, semiconductors, and data networks as core components of economic power in the twenty-first century.

Similar strategies shape policy across Europe, China, Russia, India,

Brazil, South Korea, and the Gulf states. The European Union pursues AI and fintech governance through the *Digital Services Act*, the *Digital Markets Act*, and the *EU AI Act*, each designed to regulate platform power and promote trusted AI deployment. China advances national AI capability through its *New Generation Artificial Intelligence Development Plan*, its Social Credit data architecture, and state-supported semiconductor investment. Gulf economies such as the United Arab Emirates fund sovereign AI labs and establish digital-asset-friendly regulatory regimes to attract global firms. Across the world, governments subsidize cloud and semiconductor production, expand military AI programs, and create regulatory pathways for fintech and digital assets.

Central banks now experiment with *Central Bank Digital Currencies (CBDCs)*, digitized forms of state money that can be tracked, monitored, and even programmed at the code level. These instruments promise efficiency and near-instant settlement. They also grant governments the potential to embed monetary rules directly into money itself. Public policy, once centered on manufacturing, agriculture, or trade, now treats AI, crypto, networks, and cloud architecture as strategic assets, comparable to nuclear energy or aerospace technology in earlier eras.

A defining feature of this new paradigm is marginal cost. *Marginal cost* refers to the cost of producing one additional unit. For a car or a loaf of bread, marginal cost includes steel, wheat, fuel, and labor. For a digital product, marginal cost approaches zero. Once the software, model, or platform is built, distributing one more copy, or one million more copies, requires little additional expense. Cost lies primarily in creation rather than replication. This is why a single messaging app, streaming platform, or AI-generated report can reach global audiences without factories, warehouses, or supply chains.

This dynamic produces extreme scalability. A product can serve entire continents without proportionate increases in labor or capital. A traditional manufacturer might need decades and vast investment to reach international markets. A digital platform can reach the same audience in weeks. When marginal cost approaches zero and networks reward scale, a small number of platforms tend to dominate. A search engine that is marginally better becomes the default tool for information retrieval. A payment app that gains early trust becomes the channel for millions of transactions. These platforms sit between users and the broader economy, collecting fees, shaping interaction, and accumulating data from nearly every digital action.

Digital assets intensify this shift. *Cryptocurrencies* such as Bitcoin (BTC) and Ethereum (ETH) are digital bearer instruments, meaning ownership is proven directly by possession rather than through a bank or government intermediary. Stablecoins function as digital dollars or euros that move across borders at the speed of software. Tokenization converts real-world assets, including property, securities, and artwork, into programmable entries on a blockchain. Value is no longer confined to a single jurisdiction, bank, or legal framework. It flows through decentralized or semi-decentralized networks continuously and globally.

The emerging order is a tech-centric economic system, *Techonomics*, in which value creation is global, programmable, and capable of scaling with little friction. A small team can design a protocol, an AI tool, or a digital platform that operates worldwide from its first day online. Success depends less on controlling land or machines and more on controlling code, data, and network access. Those who manage these systems command leverage far greater than most industrialists of the past.

Physical scarcity still exists and always will. Much modern wealth, however, now emerges from systems that bypass traditional constraints. Digital replication is effectively infinite. Distribution is nearly instantaneous. Value accumulates at unprecedented speed. To understand money, power, and risk in the twenty-first century, one must understand this paradigm, because the foundations of economic life have shifted.

Technology as the Core Driver of Wealth

In the *Techonomic* era, wealth no longer originates in physical scale. A company's power is not measured by the number of factories it owns, the volume of goods it ships, or the size of its workforce. Modern wealth is rooted in software, data, and global reach. The companies that dominate today's economy built their success on digital systems operating at planetary scale. Their platforms reach billions of people each day, and entire industries depend on the infrastructure they control.

A central mechanism behind this dominance is the network effect. A *network effect* occurs when a system becomes more valuable as more people use it. Google's search engine improves with every query because each interaction generates data that strengthens ranking algorithms. Facebook grew into a global forum because each new user increased the platform's value for others. Amazon's marketplace became indispensable because additional buyers and sellers improved selection, pricing, and liquidity. The result is platform dominance, in which a small group of firms captures outsized value as their systems grow stronger through use.

In this environment, wealth increasingly flows from intellectual property rather than physical production. Algorithms have become the new assembly lines. *Proprietary software*, machine learning models, *data pipelines*, and *patent portfolios* function as digital factories that produce value without consuming materials or requiring proportionate labor. A small engineering team can design systems that scale to billions of devices. Code replaces machinery. Data replaces fuel. Networks replace distribution channels.

This dynamic allows a single technological breakthrough to reshape markets with unusual speed. When Apple introduced the iPhone in 2007, it did more than launch a device. It reorganized communications, software development, photography, digital commerce, payments, and social media. Entire industries adapted around a single platform because digital systems scale in ways physical systems cannot. The same pattern is now visible in artificial intelligence. Models such as ChatGPT show how software can influence business operations, creative work, and information processing almost immediately after deployment.

Technology has become the engine of modern wealth creation rather than a tool used alongside it. It determines which firms dominate markets, which ideas spread globally, and which institutions accumulate influence. Mastery of software, ownership of data, and control of networks have become primary levers of economic power.

Data as the New Oil

In the digital economy, data has become the most valuable resource. It is often compared to oil because it can be extracted, refined, and monetized at scale. Every click, search, location check, purchase, or message creates an informational trace that can be stored, analyzed, and used to influence behavior. Data also has properties that distinguish it from physical commodities. It is reproducible and non-rivalrous, meaning the same dataset can be used repeatedly without depletion. These characteristics allow data to fuel systems often described as *surveillance capitalism*, where prediction and behavioral influence become central sources of profit.

Data gains value through compounding. Each interaction produces information that improves the algorithms relying on it. As platforms grow, their data becomes richer and more precise.

- A search engine processes billions of queries, sharpening its ability to anticipate intent.
- A video platform observes viewing patterns, refining recommendations to match individual preference.
- A social network tracks interactions and engagement, learning how to retain attention.

Over time, platforms such as Google Search or TikTok's recommendation system become more effective precisely because they have been used so extensively. Better prediction attracts more users. More users generate more data. This compounding loop gives established firms an advantage that newcomers struggle to overcome.

Artificial intelligence depends on this reservoir of data in the same way factories once depended on electricity. An AI model cannot function without training data. Large and diverse datasets allow models to recognize patterns, predict outcomes, translate languages, analyze medical images, recommend financial decisions, and generate content. Whether a system is scoring a loan applicant or producing text, it requires a continuous flow of information. Data fuels algorithms that now drive advertising, logistics, healthcare, transportation, entertainment, and administration.

This dependence has produced closed ecosystems and digital monopolies. Companies such as Meta, Google, Amazon, and ByteDance control immense data networks that are internally integrated and difficult to penetrate. Their models improve daily. Their competitive advantages deepen as their datasets expand. Even national governments struggle to match the predictive accuracy produced by continuous collection at global scale. These firms shape markets by determining visibility, access, and attention.

The implications extend beyond commerce. Control of data enables influence over public discourse, consumer behavior, financial decision-making, and cultural narratives. Data has become a strategic currency of the *Techonomic* era, shaping markets, politics, and identity. In this sense, data is not simply the new oil. It is the foundation of a parallel economy governed by those who can capture, refine, and deploy information at scale.

The Collapse of Traditional Economic Assumptions

Traditional economics rested on assumptions about labor, scarcity, and geography. Labor was viewed as the primary source of value. Scarcity defined supply. Geography constrained markets. The digital age weakens each of these foundations.

Labor was historically tied to output. Factories required workers. Farms required farmers. Offices required clerks and analysts. Today a single developer can create systems that serve millions or billions without proportional hiring. Automation and machine intelligence perform tasks that once required coordinated teams, including fraud detection, customer support, logistics management, and analysis. Productivity now arises from code that scales rapidly rather than labor that scales slowly.

Scarcity is also challenged. Digital goods such as music, video, software, and AI-generated content can be copied and distributed at almost no cost. Creating another digital file requires no additional materials or labor. When abundance becomes normal, traditional supply and demand models lose explanatory power. Value shifts from ownership of goods to control of platforms, algorithms, and attention.

Economic measurement struggles to keep pace. Metrics such as GDP were designed for an industrial economy centered on physical output. They do not fully capture the value created by *open-source communities*, global platforms, cloud infrastructure, recommendation systems, or generative AI. Systems that reshape communication and behavior often register only marginally in official statistics.

Geography has also lost much of its force. Digital goods cross borders instantly. Markets become global by default. Success depends less on location and more on software quality, network scale, and algorithmic effectiveness.

These changes reshape how wealth concentrates. In the industrial world, power flowed through factories, land, and raw materials. In the digital world, power flows through platforms, data pipelines, cloud systems, and AI models. The infrastructure that matters most is no longer steel or concrete. It is digital architecture. Whoever controls the platforms increasingly controls the markets built on top of them.

The conclusion follows naturally. The old rules no longer define the new environment. Labor is no longer the primary constraint. Scarcity is no longer universal. Borders no longer limit competition in the same way. Networks, data, and algorithms have rewritten the economic physics of the modern age.

Technocrats

Traditional economics shaped societies for centuries by linking wealth to tangible resources. Production expanded through workers, factories, and machines. Banks, industrial conglomerates, and governments shaped economic life because they controlled these assets.

Techonomics rests on a different foundation. Once an AI model, dataset, or digital platform exists, it can be deployed globally at minimal cost. Firms rise to dominance by writing the code that other institutions depend upon. Power flows toward those who design and govern systems that cannot easily be replaced.

This shift has produced a new technocratic class. In the industrial age, influence belonged to oil magnates, railroad owners, steel barons, and financiers. In the digital age, influence belongs to founders, executives, and engineers who oversee platforms that underpin communication, commerce, and security. Figures associated with transportation, logistics, cloud infrastructure, social networks, and AI hardware do not merely run companies. They manage systems that function as essential infrastructure for modern life.

The rise of this class carries cultural and political consequences. Pope Francis warned of a growing technocratic paradigm in his 2015 encyclical Laudato Si'. His concern was not technology itself, but the belief that technical capability automatically produces human progress. Such a worldview elevates efficiency over meaning and optimization over wisdom, often displacing moral judgment and charity with managerial logic.

Political conflict once centered on ideology. Today influence often flows through tools rather than doctrine. Platforms determine visibility. Algorithms shape belief. Data pipelines influence markets, elections, and culture. The public square is increasingly mediated by private systems operated by a small number of technical elites.

A quiet transition has taken place. Twentieth-century societies believed military strength and industrial output defined power. Yet governments and corporations constructed technological systems that now overshadow older institutions. Scientific innovation brings genuine benefit. It also concentrates authority among those who command the computational, financial, and data-driven machinery of modern life.

By the early 2020s, this consolidation became visible. Political actors relied on technology firms for capital, infrastructure, and influence. Technology firms relied on government contracts and regulatory alignment to preserve dominance. Allegiances shifted with political conditions because the underlying incentive was preservation of technical and economic authority rather than ideological commitment.

The technocratic worldview is utopian in form. It imagines automation, digital currency, and artificial intelligence eliminating the burdens of work and allowing society to be managed scientifically. In that vision, technocrats become architects of social order. The risk is not malice, but substitution, when optimization replaces judgment.

Societies and regulators remain poorly prepared for this reality. Tax systems, antitrust frameworks, and trade laws were designed for factories and physical scarcity, not for platforms, algorithms, and global data flows. Without new approaches to governance, concentration of influence among a small group of technological actors will continue.

Techonomics is not an extension of *classical economics*. It is a different paradigm. To navigate it, societies will need clearer measures of value, stronger political tools to manage digital power, and a deeper understanding of how the technocratic class shapes the modern world. Those who rely on industrial assumptions will find themselves acting inside a system that no longer answers to the old rules.

Chapter III

Digital Assets

Blockchain Basics

The modern digital asset economy began in 2009 with the release of Bitcoin (BTC), the first form of money to exist entirely online without relying on banks, corporations, or governments. It was introduced by an anonymous creator known as Satoshi Nakamoto, who in October 2008 published a white paper titled Bitcoin: A Peer-to-Peer Electronic Cash System. The *Bitcoin white paper* described a method for creating digital money that could not be forged, censored, or manipulated by a central authority. In January 2009, the Bitcoin (BTC) network launched. For the first time, a financial system operated through decentralized software rather than through centralized institutions.

At the center of this system is the *blockchain*, the technology that makes decentralized money possible. A blockchain is a public digital ledger that records transactions in a permanent, chronological sequence. Unlike a bank ledger controlled by one institution, a blockchain is maintained by thousands of independent computers around the world. Anyone can view its activity through free online tools known as *block explorers*. This transparency makes covert alteration difficult to hide. Once information is recorded on a blockchain and sufficiently confirmed by the network, it becomes effectively immutable, which is one reason fraud is so difficult.

Because blockchains are decentralized networks rather than companies or apps, each major cryptocurrency operates its own chain. Bitcoin (BTC) has the Bitcoin blockchain. Ethereum (ETH) has the Ethereum blockchain. Other digital assets either run on their own networks or exist as tokens built on top of another blockchain. Thousands of tokens, for example, run on the Ethereum (ETH) blockchain without requiring their own independent infrastructure.

To keep these networks secure, blockchains rely on *cryptography*, the science of encoding information so that only authorized parties can access or verify it. Every transaction uses a *private key*, a secret digital signature proving that the sender controls the funds being transferred. The network verifies this signature using a corresponding *public key*. These keys are linked through mathematical methods that make ownership easy to prove but practically impossible to counterfeit. This system, known as *asymmetric cryptography*, is the same general method used in secure communications, online banking, and military encryption. Because the underlying mathematics is computationally infeasible to reverse, no one can forge ownership or rewrite a transaction after it has been recorded.

As transactions accumulate, blockchains store them in groups called *blocks*, which are added one after another to form a chain. This structure requires a way for thousands of independent computers to agree on which transactions are valid and which block comes next. To solve this, blockchains use consensus mechanisms. A *consensus mechanism* is a rule system that allows the network to agree on a single version of the ledger without a central authority deciding for everyone.

Consensus rules are not improvised after launch. They are chosen by a blockchain's founders before the network goes live. When a new blockchain is designed, its creators choose which consensus system it will use, how new blocks will be added, how participants will be rewarded, and how the network will remain secure over time. These rules are embedded directly into the blockchain's software. Once live, every *node* follows the same rules, and no single participant can change them unilaterally. Any update requires broad, voluntary agreement across the network's participants.

Two major consensus systems dominate today.

1. **Proof of Work (PoW)**

This system requires miners to compete to solve mathematical problems using specialized computers. The first miner to solve the puzzle earns the right to add the next block and receives newly created cryptocurrency as a reward. Bitcoin (BTC) uses *Proof of Work (PoW)*. It is highly secure because cheating would require enormous computing power, though it also consumes significant electricity.

2. **Proof of Stake (PoS)**

   This system selects validators based on how much cryptocurrency they lock up as collateral. Validators verify transactions and create new blocks. If a validator behaves dishonestly, a portion of their collateral can be removed. *Proof of Stake (PoS)* uses far less energy and can support greater scalability. Ethereum (ETH) transitioned to this system in 2022.

Bitcoin (BTC) and Ethereum (ETH) illustrate how these approaches differ. Bitcoin (BTC) continues to rely on Proof of Work, prioritizing durability and decentralization even at higher energy cost. Ethereum (ETH) moved from Proof of Work to Proof of Stake to reduce energy consumption and support a broader range of applications, including decentralized finance (DeFi) and smart contracts. Bitcoin emphasizes simplicity and permanence. Ethereum emphasizes flexibility and ecosystem growth.

Staking allows cryptocurrency holders to support a Proof of Stake blockchain by locking up coins as collateral in return for rewards. Instead of relying on physical mining hardware, staking relies on economic commitment and operational reliability. Networks such as Ethereum (ETH), Cardano (ADA), Solana (SOL), Avalanche (AVAX), and Polkadot (DOT) select validators based on how much they have staked and how consistently they perform. Users who do not wish to run validator infrastructure can delegate their stake to trusted validators through wallets or exchanges, earning a share of rewards without managing the underlying systems.

Staking is generally limited to *Layer 1* coins because these assets directly secure their base blockchains. *Layer 2* networks such as Optimism (OP) and Arbitrum (ARB) may have governance tokens, but they do not use staking to secure the Ethereum (ETH) base layer. Most application tokens and stablecoins, including Chainlink (LINK), Uniswap (UNI), USD Coin (USDC), and Tether (USDT), cannot be staked for network security because they do not run their own blockchains. In Proof of Stake systems, staking is therefore primarily a feature of native Layer 1 coins.

Because blockchain networks vary in purpose, they can be either public or private. A *public blockchain* is open to anyone. Anyone can download the software, create a wallet, send transactions, or examine the transaction history. Bitcoin (BTC) and Ethereum (ETH) are the best-known public blockchains because they are permissionless and globally accessible.

A *private blockchain* follows a different model. It is a permissioned network where only approved users can participate. Access to data, transaction submission, and node operation is restricted. This allows organizations to benefit from blockchain properties such as immutability and shared verification while keeping sensitive information confidential. Banks, corporations, and government agencies use private blockchains when they need secure coordination across multiple institutions without exposing internal data publicly. Systems such as Hyperledger Fabric, R3 Corda, and Quorum provide this controlled environment.

Industries adopt these systems for specific use cases. Hyperledger Fabric, developed by the Linux Foundation and International Business Machines Corporation (IBM), is used by retailers such as Walmart to track food safety and product origins across supply chains. Shipping and logistics firms use Fabric to unify records that normally remain split across separate databases. R3 Corda, created by the financial technology firm R3, is used by institutions including Hongkong and Shanghai Banking Corporation (HSBC), Banque Nationale de Paris Paribas (BNP Paribas), and Internationale Nederlanden Groep (ING) to settle trades, share audit information, and reduce reconciliation errors. Quorum, originally developed by JPMorgan as an enterprise version of Ethereum (ETH), is used by central banks and government agencies exploring digital currency pilots. Healthcare providers and public institutions use private blockchains to secure medical records, identity systems, and audit trails without exposing sensitive data. In each case, private blockchains offer shared verification inside a controlled membership.

Public and private networks also differ in how they use cryptocurrency. Every public blockchain requires a native coin to operate. Bitcoin (BTC) is the native coin of the Bitcoin blockchain. Ethereum (ETH) is the native coin of the Ethereum blockchain. Native coins pay transaction fees, reward miners or validators, and support the functions that keep a public network secure. They are essential because participants in public networks can be anonymous and must be incentivized to act honestly.

Private blockchains operate differently. Because participation is restricted to known and authorized members, they do not rely on mining rewards, staking systems, or transaction fees paid in a native coin. Hyperledger Fabric, R3 Corda, and Quorum have no native cryptocurrency at all. Their security comes from governance, identity controls, and organizational trust rather than economic incentives.

As the ecosystem expanded, developers introduced layers of infrastructure. A Layer 1 network is a base blockchain that operates its own ledger and security system. Bitcoin (BTC), Ethereum (ETH), Solana (SOL),

Cardano (ADA), and Avalanche (AVAX) are examples. Each uses its native coin to pay network fees and support its consensus mechanism. These Layer 1 networks form the foundation of the digital asset landscape.

A Layer 2 network is designed to scale a Layer 1 blockchain. It processes transactions off the base chain and then settles results back onto it. The Lightning Network is the primary Layer 2 for Bitcoin (BTC), enabling fast, low-cost payments. Ethereum (ETH) supports several Layer 2 rollups such as Optimism (OP), Arbitrum (ARB), Base, and zkSync (ZK), which bundle large volumes of transactions off-chain and submit compressed proofs to Ethereum (ETH) for settlement. By moving most activity off the congested base layer, Layer 2 systems improve speed and reduce costs while retaining the base layer's security model.

Tokens represent a more flexible category of digital assets. Unlike coins, tokens do not have their own blockchains and do not secure or scale a network. They are created through smart contracts on existing blockchains. A coin belongs to a blockchain itself. A token belongs to an application or protocol running on that blockchain. Tokens can represent utility, governance, access rights, or value within a digital ecosystem.

Most blockchain ecosystems host their own set of tokens. On Ethereum (ETH) Layer 1, notable examples include Chainlink (LINK), Uniswap (UNI), Aave (AAVE), and The Sandbox (SAND), all of which rely on Ethereum for settlement and security. *Ethereum (ETH) Layer 2* networks issue governance tokens such as Optimism (OP), Arbitrum (ARB), and zkSync (ZK), while Base currently operates without a native token.

Solana (SOL) Layer 1 hosts application tokens such as Raydium (RAY) and STEPN (GMT), reflecting it's high-thoughtput, low-latency design. Earlier Solana-era protocols such as Serum (SRM) played a formative role in the ecosystem but declined following *the FTX scandal* in late 2022. *Cardano (ADA) Layer 1* supports tokens such as SingularityNET (AGIX), which operates across multiple chains, and MELD (MELD). *Avalanche (AVAX) Layer 1* hosts decentralized applications like Trader Joe (JOE) and Pangolin (PNG), as well as subnet-based ecosystems including DeFi Kingdoms (JEWEL) and Shrapnel (SHRAP). Although diverse, all tokens depend on their underlying blockchain or Layer 2 for security and operation.

Stablecoins form a distinct class of tokens designed to track the value of fiat currencies such as the US dollar. Major stablecoins include USD Coin (USDC), Tether (USDT), and Dai (DAI). *ERC-20* is a standardized Ethereum (ETH) token format that defines how *fungible digital assets* are issued, transferred, and accounted for on the Ethereum (ETH) Layer 1 blockchain. Many of these tokens originate on Ethereum (ETH) Layer 1 but also circulate through bridged representations on Ethereum (ETH) Layer 2 networks such as Optimism (OP), Arbitrum (ARB), and Base, as well as on Layer 1 blockchains including Solana (SOL), Avalanche (AVAX), and Polygon (MATIC). Stablecoins do not secure or scale a blockchain, but they have become vital for payments, trading, lending, and decentralized finance.

Users interact with blockchains through *digital wallets* such as MetaMask, Coinbase Wallet, or hardware devices like Ledger. A wallet does not store cryptocurrency directly. It stores private keys that prove ownership of cryptocurrency recorded on the blockchain. When a user sends funds, the wallet signs the transaction with the private key and broadcasts it to the network. Miners or validators confirm it and record it.

Because blockchains record ownership transparently and immutably, users can verify assets, track transaction history, and transfer value anywhere in the world without relying on banks, escrow services, or clearinghouses. This removes layers of cost and bureaucracy that have long shaped traditional financial systems.

In little more than a decade, blockchains have become infrastructure for cryptocurrencies, stablecoins, digital identity systems, NFTs, decentralized finance, synthetic assets, and early forms of government-backed digital currencies. They create a world where trust is generated through transparent mathematics rather than institutional authority. For the first time in modern history, financial systems can operate without central control, and ownership can be verified by software rather than by intermediaries.

Brief History of Digital Assets

Digital assets began in 2009 with the creation of Bitcoin (BTC), a *Layer 1 blockchain* released by the anonymous developer Satoshi Nakamoto. Although Nakamoto's identity remains unknown, their writings and early communications showed deep expertise in cryptography, computer science, and economics. Bitcoin (BTC) was the first currency to exist entirely online without control from any bank or government. It demonstrated that money could move directly between individuals across the internet with no intermediary. From 2009 through 2012, Bitcoin (BTC) was used mainly by programmers and early adopters who believed cryptographic money could provide financial independence. As adoption grew, Bitcoin (BTC) became known as digital gold.

What made Bitcoin (BTC) especially significant was the moment in which it appeared. Nakamoto released the Bitcoin white paper in October 2008, just weeks after the collapse of major financial institutions and the beginning of historic government bailouts. The Bitcoin network launched on January 3, 2009, the same day The Times published the headline "Chancellor on brink of second bailout for banks." Nakamoto encoded that headline directly into Bitcoin's first block, known as the *Genesis Block*. It served both as a timestamp and as a critique of the traditional financial system, which had revealed vulnerabilities in transparency, trust, and monetary stability. Bitcoin (BTC) was designed as an alternative: a currency with a fixed supply, no central authority, and a public ledger that no institution could quietly alter.

The 2008 crisis exposed how dependent society had become on centralized institutions that could fail, be rescued, or expand the money supply without clear limits. Bitcoin (BTC) offered a different model rooted in decentralization and mathematical assurance. Its supply cap of 21 million coins restrained inflation by design, and its peer-to-peer architecture removed the need for banks to approve or process transactions. These features attracted people who saw Bitcoin (BTC) not only as a technological breakthrough but also as a hedge against institutional risk and monetary debasement. That philosophy helped shape the next wave of digital assets as developers began exploring how blockchains, tokens, and smart contracts could extend Bitcoin's logic into broader financial systems.

By 2011 and 2012, new cryptocurrencies appeared. These were soon called *altcoins*, meaning assets that were not Bitcoin (BTC). Litecoin (LTC), a Layer 1 blockchain launched in 2011, was created by Charlie Lee, a former Google engineer who later became Director of Engineering at Coinbase. Litecoin (LTC) offered faster confirmation times and became one of the earliest alternatives to Bitcoin (BTC). XRP (XRP), the native coin of the XRP Ledger, a Layer 1 blockchain launched in 2012, was developed by Arthur Britto, David Schwartz, and Jed McCaleb. McCaleb had previously founded Mt. Gox, one of the first Bitcoin (BTC) exchanges. XRP (XRP) was later promoted by Ripple, the American financial technology company founded by Chris Larsen and Jed McCaleb, as a cross-border payments tool used in pilots and integrations with banks and financial institutions. Many early altcoins focused on speed, efficiency, and energy usage.

A major turning point came in 2015 with the launch of Ethereum (ETH), a Layer 1 programmable blockchain created by Vitalik Buterin, a Russian-Canadian programmer who previously cofounded Bitcoin Magazine. Ethereum (ETH) was developed by the Ethereum Foundation, a Switzerland-based nonprofit. Ethereum (ETH) introduced smart contracts, allowing developers to build applications directly on a blockchain. After Ethereum's release, new Layer 1 networks emerged with specialized aims.

Cardano (ADA), launched by Input Output Global under Charles Hoskinson, a former Ethereum (ETH) cofounder, emphasized peer-reviewed academic research. Solana (SOL), launched by Solana Labs under former Qualcomm engineer Anatoly Yakovenko, focused on high-speed processing. Polkadot (DOT), created by Gavin Wood, another Ethereum (ETH) cofounder and creator of the Solidity programming language, aimed to connect multiple blockchains through interoperability.

Digital Assets

New categories formed as developers pursued different tradeoffs. *Privacy blockchains* are networks designed so that transaction details, wallet addresses, and account balances remain hidden, allowing users to transact without exposing financial activity to the public. Privacy assets such as Monero (XMR), a Layer 1 privacy blockchain launched in 2014 by a group of developers associated with Riccardo Spagni, and Zcash (ZEC), a Layer 1 privacy blockchain launched in 2016 by the Electric Coin Company under cybersecurity researcher Zooko Wilcox, used advanced cryptography to conceal transaction details. Cross-chain systems also appeared. Wrapped Bitcoin (WBTC), an *ERC-20* token on Ethereum (ETH) launched in 2019 by BitGo, Kyber, and Ren, allowed Bitcoin (BTC) to be used in Ethereum-based applications without converting it into a different base asset.

Stablecoins became one of the most consequential innovations. Tether (USDT), a stablecoin launched on Ethereum in 2014 by Tether Limited founders Brock Pierce, Reeve Collins, and Craig Sellars, created the first widely used digital dollar proxy. USD Coin (USDC), a stablecoin launched in 2018 by Circle and Coinbase through the Centre Consortium, offered a more regulated dollar-backed structure with reserve disclosures. Dai (DAI), a decentralized, crypto-collateralized stablecoin launched in 2017 by MakerDAO under the leadership of Rune Christensen, pursued stability without relying on banks by using on-chain collateral and rules-based mechanisms. Stablecoins reduced volatility and made continuous trading, lending, and cross-border transfers more practical.

Beginning in 2017, the digital asset universe expanded quickly. Chainlink (LINK), a token launched in 2017 by Sergey Nazarov and Chainlink Labs, provided *decentralized oracles* that deliver external data to smart contracts. Uniswap (UNI), a governance token launched in 2020 by Hayden Adams, enabled token holders to vote on changes to the Uniswap protocol. These *governance tokens* introduced a new structure for decentralized decision-making.

At the same time, *Non-Fungible Tokens (NFTs)* emerged. CryptoPunks, *ERC-721* NFTs launched in 2017 by Matt Hall and John Watkinson of Larva Labs, demonstrated that digital images could be uniquely owned through blockchain verification. Bored Ape Yacht Club, ERC-721 NFTs launched in 2021 by Yuga Labs founders Greg Solano and Wylie Aronow, pushed NFTs into mainstream pop culture through entertainment, branding, and celebrity adoption. Over time, NFTs expanded beyond collectibles into identity, licensing, gaming assets, ticketing systems, and membership access.

Synthetic assets and *tokenization* grew alongside these developments. Synthetix (SNX), a token launched in 2018 by Kain Warwick, enabled synthetic versions of real-world assets including stocks, commodities, and fiat currencies. Tokenization brought real estate, gold, treasury bills, corporate shares, and other assets onto blockchains in fractional digital form, reshaping liquidity and settlement expectations.

Digital asset gaming became another major sector. Decentraland (MANA), a token launched in 2017 by Esteban Ordano and Ari Meilich, and The Sandbox (SAND), a token relaunched in 2018 by Arthur Madrid and Sebastien Borget at Animoca Brands, created virtual worlds where users could own digital land, build environments, and trade virtual items. These spaces evolved into token-driven economies where ownership and commerce were recorded on-chain.

Between 2018 and 2019, *cross-chain* and *AI-connected tokens* gained visibility. Fetch.ai (FET), a token launched in 2019 on Cosmos and Ethereum (ETH) by Humayun Sheikh, Toby Simpson, and Thomas Hain, explored autonomous AI agents operating on decentralized networks. SingularityNET (AGIX), launched in 2017 by AI researcher Ben Goertzel, sought to create a decentralized marketplace for artificial intelligence services across chains, including Ethereum and later Cardano.

From 2019 through 2021, *decentralized finance (DeFi)* reshaped digital markets. *Lending pools*, *automated market makers*, *derivatives protocols*, and *yield strategies* introduced tokens that represented financial rights within decentralized systems. Environmental and carbon-focused tokens expanded the ecosystem further and aligned blockchain tools with sustainability initiatives.

Government involvement grew around 2020 with the expansion of *central bank digital currencies (CBDCs)*. The Digital Yuan, a state-issued CBDC on a permissioned system launched by the People's Bank of China, became the first large-scale example of a government-controlled digital currency. *Regenerative Finance (ReFi)* also emerged, with projects such as KlimaDAO (KLIMA), a token launched in 2021 on the Polygon network, aimed at supporting carbon offset markets and climate initiatives.

Alongside these developments, a different class of digital assets gained prominence that did not derive value from protocol necessity or institutional design. So-called *memecoins* and *socially driven tokens* emerged without governing infrastructure, required utility, or systemic function. Their value rested instead on narrative coordination, speculation, and collective attention. Unlike assets such as Bitcoin, Ethereum, or governance tokens that secure networks or allocate decision rights, these tokens operated primarily as cultural instruments, rallying liquidity through humor, identity, and momentum rather than through technical dependence. Markets nonetheless priced them aggressively, demonstrating that belief, coordination, and velocity could generate real financial outcomes even in the absence of functional necessity. This divergence revealed a deeper truth of the *Techonomic* era: digital markets do not reward utility alone, but also the capacity to mobilize attention at scale.

Institutional participation followed. Products such as Grayscale Bitcoin Trust (GBTC), created by Grayscale Investments under Digital Currency Group, allowed traditional investors to gain exposure to Bitcoin (BTC) without holding it directly. GBTC launched as a statutory trust at a time when Bitcoin exchange traded funds were not yet permitted in U.S. markets, which made it one of the first widely accessible Bitcoin investment vehicles for brokerage and retirement accounts. After federal approval of spot Bitcoin ETFs in 2024, GBTC was converted from a trust into a regulated *exchange traded fund (ETF)*, allowing it to trade alongside conventional ETFs. In parallel, financial institutions explored tokenization frameworks, institutional custody services, and blockchain-based settlement systems.

Today the digital asset landscape is broad and varied. It includes Layer 1 coins, Layer 2 tokens, governance tokens, staking assets, stablecoins, synthetic instruments, wrapped tokens, NFTs, virtual property, tokenized real-world assets, AI-connected tokens, and government-issued digital currencies. In little more than a decade, digital assets have evolved from a single experimental idea into a global financial architecture that shapes how people store value, transfer money, own digital property, and participate in twenty-four-hour global markets.

Decentralized Finance (DeFi)

Decentralized Finance (DeFi) refers to financial services that operate on public blockchains rather than through banks, brokerages, or traditional intermediaries. In this environment, software replaces gatekeepers. Anyone with an internet connection can access financial tools that execute automatically through transparent code. The result is a parallel financial system that is global, permissionless, and continuously operational.

The foundation of DeFi is the *smart contract*, a self-executing program that performs a defined action once specific conditions are met. Because smart contracts operate without human approval, DeFi applications become predictable and accessible regardless of geography or institutional status. The rules are visible in advance, execution is automatic, and participation does not depend on credit history or regulatory permission.

Governance tokens and protocol tokens guide how these decentralized systems evolve. *Governance tokens* allow holders to vote on changes to a protocol, including fee structures, risk parameters, upgrades, and treasury use. For example, Aave (AAVE), an Ethereum Layer 1 and Layer 2 governance and utility token, and Compound (COMP), an Ethereum Layer 1 governance token, allow holders to influence lending limits and collateral requirements. *Protocol tokens* also serve functional roles within applications by rewarding participation, incentivizing liquidity, and granting access to features. These tokens do not secure the underlying blockchain. They coordinate behavior within individual platforms.

One major category within DeFi is the *lending protocol.* These systems allow users to lend or borrow cryptocurrency without credit checks or bank approval. Platforms such as Aave (AAVE) and Compound (COMP) pool user deposits and permit borrowing only against collateral that exceeds the value of the loan. This overcollateralization protects the system from default and allows the protocol to manage risk automatically. If collateral value falls below required thresholds, smart contracts liquidate positions without human intervention.

Trading operates differently as well. *Decentralized exchanges (DEXs)* allow users to trade assets directly through smart contracts rather than through a centralized exchange. Prices are determined by algorithms that monitor the ratio of assets in *liquidity pools.* Users who supply liquidity earn fees in return for providing capital. DEXs such as Uniswap (UNI), an Ethereum Layer 1 and Layer 2 governance token, and Curve (CRV), an Ethereum Layer 1 and Layer 2 governance and incentive token, operate continuously because they rely on automated execution rather than market hours or centralized order books.

DeFi relies heavily on stablecoins, which function as digital cash within these systems. Stablecoins such as USD Coin (USDC), Tether

(USDT), and Dai (DAI) maintain relatively stable prices while moving at blockchain speed. They make borrowing, lending, and trading practical by reducing volatility. Because they behave like currency but operate without banks, stablecoins serve as the settlement layer of the DeFi ecosystem.

Yield markets emerged as another defining feature. In these systems, users earn returns by depositing assets into pools that support lending, trading, or other protocol functions. Returns may come from interest, trading fees, or newly issued tokens. Yield is generated entirely through automated mechanisms rather than discretionary management. This creates a form of passive participation that depends on code performance rather than institutional trust.

Some platforms also support *on-chain credit*, where borrowing and lending occur entirely through smart contracts. There are no applications, approvals, or human oversight. Users borrow against digital collateral, and smart contracts adjust or liquidate positions automatically when conditions change. Risk management is enforced mathematically rather than administratively.

The core principle of DeFi is direct participation. Financial services no longer require centralized intermediaries. Instead, they operate on public infrastructure where the rules are visible, execution is automatic, and access is open. DeFi demonstrates how money, credit, and markets can function when trust is placed in transparent software rather than institutional authority. It reveals both the efficiency and the fragility of systems governed by code, setting the stage for broader debates about risk, governance, and responsibility in the *Techonomic* era.

Smart Contracts

Smart contracts are computer programs stored on a blockchain that automatically execute an agreement once specific conditions are met. They do not rely on lawyers, courts, or human enforcement. Their logic is fixed in code, and execution occurs deterministically. If the conditions are satisfied, the outcome follows. Because the code runs across thousands of decentralized nodes, no single party can alter its behavior or halt its execution. As long as the blockchain remains operational, the contract performs exactly as written.

A simple comparison is a vending machine. When the correct payment is inserted, the machine releases the product without discretion or negotiation. A smart contract functions the same way, but in digital form. Payment, verification, and settlement occur automatically. This shifts enforcement from institutions and judgment to mathematics and software. Trust is no longer placed in an intermediary, but in the transparency and reliability of the code itself.

The Ethereum (ETH) blockchain, launched in 2015 by Vitalik Buterin and a team of developers, made smart contracts widely accessible.

Earlier blockchains focused primarily on payments. Ethereum (ETH) was designed as a programmable network, allowing developers to build *decentralized applications*, commonly called *dApps*, directly on the blockchain. These applications run continuously and cannot be altered unilaterally once deployed.

A basic example is a blockchain based savings application. A user deposits funds, and the smart contract automatically calculates and distributes interest without a bank managing the account. Other dApps handle ownership records, payment processing, financial markets, digital identity, governance systems, and digital goods. This design transformed Ethereum (ETH) into foundational infrastructure for decentralized finance, NFT markets, gaming economies, and a wide range of blockchain based platforms.

To interact with real world information, smart contracts rely on *oracles*. A blockchain cannot directly access external data, so an oracle supplies verified information such as asset prices, weather conditions, shipment confirmations, or event outcomes. A lending protocol, for example, requires accurate price feeds to manage collateral. An insurance contract may need confirmation that a flight was delayed or a shipment was lost. Oracles act as data bridges between blockchains and the external world. Chainlink (LINK), a token on Ethereum Layer 1, is the most widely used decentralized oracle network.

Smart contracts support a wide range of practical uses. In supply chains, they verify delivery milestones and release payments only when conditions are met. In insurance, they automate payouts based on predefined triggers. In financial markets, they manage lending, borrowing, trading, and collateral without human oversight. They also govern NFT ownership, ticketing systems, loyalty programs, digital identity, and in game economies. Any activity that depends on verification, conditional execution, or trust can be automated through smart contract logic.

A more advanced structure built from smart contracts is the *Decentralized Autonomous Organization (DAO)*. A DAO is an organization governed by rules encoded on a blockchain rather than by executives or boards. Token holders submit proposals and vote on decisions. Once a vote concludes, smart contracts automatically execute the outcome. Governance becomes a transparent, rule-driven process in which authority is distributed among participants rather than concentrated in leadership.

Smart contracts also introduce a tension between algorithmic enforcement and legal interpretation. Traditional contracts allow courts to weigh context, intent, and ambiguity. Smart contracts remove ambiguity by enforcing rules exactly as written. This produces speed, predictability, and lower costs, but it also creates risk. If a contract contains a flaw or unintended loophole, it may execute incorrectly, and reversing a blockchain transaction can be extremely difficult. These challenges have sparked global debate about how legal systems should recognize, regulate, or integrate blockchain based agreements.

Smart contracts represent one of the most consequential innovations in digital finance and governance. They convert trust into mathematics, replace discretion with execution, and shift authority away from institutions toward decentralized networks. By embedding rules directly into software, they redefine how agreements, markets, and organizations can function in a digital economy. Their continued expansion suggests a future in which economic and social coordination increasingly depends on programmable systems that operate across borders, minimize human intervention, and enforce rules through transparent logic rather than institutional power.

Crypto as a Hedge

Bitcoin (BTC) is often described as digital gold, a store of value in an era when traditional currencies can be expanded without formal limit. Its fixed supply of twenty-one million units is enforced by software rather than by policy. No central bank can increase it, and no government can alter it without changing the underlying rules of the network. This constraint gives Bitcoin (BTC) a monetary character unlike fiat currencies, whose supply can be adjusted in response to political or economic pressure. For this reason,

Bitcoin (BTC) functions as a hedge against monetary expansion, fiscal instability, and the long-term erosion of purchasing power.

Bitcoin (BTC) also introduces a new form of financial sovereignty. Ownership is proven by cryptographic keys rather than by institutional records. Funds can be held without custodians, transferred without permission, and stored without reliance on banks or clearinghouses. Historically, this level of autonomy was available only through physical assets such as gold or land. Bitcoin (BTC) offers similar independence in digital form, with the added benefits of portability, divisibility, and global transferability.

Stablecoins play a different but equally important role within the digital asset ecosystem. Stablecoins such as USD Coin (USDC), Tether (USDT), and Dai (DAI) are designed to maintain relatively stable value while moving across blockchain networks at the speed of software. They function as digital dollars within crypto markets, enabling payments, lending, and settlement without exposure to extreme price volatility. Because stablecoins operate continuously and across borders, they allow commerce and remittances to occur without banks, wire transfers, or settlement delays.

In regions with unstable currencies, capital controls, or limited banking access, stablecoins have already become practical financial tools rather than speculative instruments. They allow individuals and businesses to preserve value, receive income, and conduct trade in a currency that is not tied to local monetary instability. In this context, stablecoins act as a hedge against currency risk and institutional fragility rather than against inflation alone.

Tokenization refers to the process of translating ownership rights in a real-world or financial asset into digital representations recorded on a blockchain. At its core, tokenization does not eliminate the underlying asset. It changes how claims on that asset are issued, tracked, and transferred. A property, bond, commodity, or revenue stream remains legally anchored in the physical or contractual world, but its ownership interests are divided into digital tokens that function as records of entitlement. Each token represents a defined claim, such as equity, yield rights, or usage rights, enforced through a combination of legal agreements and smart contracts.

In practice, tokenizing real estate begins with a legal structure that holds the property, often a special purpose entity. Ownership in that entity is then digitized into tokens, which are issued on a blockchain and mapped to the legal claims defined off-chain. Smart contracts govern how tokens can be transferred, how income is distributed, and how compliance requirements are enforced. Investors can buy fractional interests, trade them globally, and receive automated payouts without relying on traditional brokers, registrars, or settlement systems. The result is that assets once limited by geography, illiquidity, and administrative friction become divisible, programmable, and accessible to a wider market, while still remaining tied to real economic value.

Non fungible tokens extend this principle to unique digital property. An NFT is not merely a digital image. It is a cryptographic record of ownership stored on a blockchain. This record can represent art, licenses, access rights, identity credentials, or intellectual property. Because ownership is verifiable and transferable without intermediaries, NFTs provide a framework for digital scarcity and authenticity in environments where duplication was previously unavoidable.

Taken together, these developments show that crypto is not simply a new asset class. It is an alternative financial architecture. Blockchains provide the base layer. Smart contracts automate enforcement. Tokens represent value, access, and ownership. Decentralized finance builds markets that operate without banks or brokers. Each component reduces dependence on centralized institutions and replaces administrative control with transparent rules encoded in software.

In this sense, crypto functions as a hedge not only against inflation or market volatility, but against the concentration of financial power itself. Traditional systems rely on intermediaries that can freeze accounts, reverse transactions, or restrict access. Blockchain based systems allow individuals to hold assets directly, transact without permission, and participate in markets governed by visible logic rather than opaque policy.

Techonomics

Crypto introduces a shift in how value is stored, transferred, and governed. Property rights, financial exchange, and even identity can now be embedded in software rather than administered exclusively by institutions. For those who learn to navigate this environment, digital assets offer tools for resilience, autonomy, and global participation in a rapidly digitizing economy. For those who ignore it, the risk is not merely missed opportunity but increasing dependence on systems whose rules they do not control.

Chapter IV

The AI Mind

How AI Works

Artificial Intelligence (AI) is not a robot or a mysterious machine. It is a mathematical prediction system, a form of software that takes inputs and produces the most statistically likely output based on patterns it has learned. At its core, AI is a prediction engine. After being trained on millions or billions of examples, it can generate responses or decisions with a speed, scale, and consistency no human being could match. AI does not possess intuition, conscience, or wisdom. It performs pattern recognition at industrial scale.

AI differs fundamentally from general intelligence, which is the flexible, contextual reasoning human beings use every day. A child can learn a lesson in one situation and apply it to a completely different one. Humans interpret emotion, recognize sarcasm, notice contradictions, and understand when rules require exceptions. AI cannot do this. It is narrow intelligence, meaning it can perform only the specific task it was trained to perform. A language model cannot drive a vehicle. A chess engine cannot diagnose a medical scan. A navigation algorithm cannot write a legal contract. Each AI system operates within boundaries defined entirely by the data it was given.

One of the most important differences between traditional software and AI lies in how rules are formed. Traditional software follows explicitly written instructions created by human programmers. AI, by contrast, learns its rules indirectly. These rules emerge internally as the model processes vast datasets and identifies patterns, correlations, and probabilities. Developers do not specify every condition by hand. They provide examples, and the system constructs its own internal logic. For this reason, AI is described as data-driven. The quality, diversity, and scale of the training data largely determine the capability of the final model.

The foundational structure behind AI is the *neural network*, a layered mathematical framework inspired loosely by the human brain. A neural network consists of many interconnected nodes that process information step by step. Early layers detect simple features such as edges, colors, or basic sounds. Middle layers detect shapes, relationships, or sentence structure. Deeper layers extract abstract meaning. A small network can recognize handwritten digits. A large network can generate lifelike images, analyze audio, write extended essays, or predict market behavior. Neural networks allow AI systems to perform pattern recognition at levels traditional programming cannot reach.

AI systems improve through a process known as training. During training, the model is exposed to large datasets and adjusts its internal parameters to reduce errors. When this process is adapted for a specific domain, it becomes fine-tuning, which uses targeted data for areas such as medicine, finance, or law. Once training is complete, the model performs inference. *Inference* is the stage where the system applies what it has learned to interpret or generate information it has never encountered before.

What makes AI transformative is its ability to scale learning without fatigue. Humans tire, forget, and reach cognitive limits. AI does not. It can analyze millions of documents in seconds. It can detect patterns across global datasets. It improves as more data flows through it. This creates a compounding effect. More users generate more data. More data improves the model. Better models attract more users. This feedback loop explains why AI has spread so rapidly into nearly every sector of the modern economy.

AI systems increasingly replace repetitive judgment with statistical precision. Recommendation engines, fraud detection systems, medical triage tools, logistics routing software, autonomous vehicles, investment models, and planning systems all rely on predictive algorithms. AI evaluates probability rather than moral meaning. It is not magic. It is mathematics applied at industrial scale, shaping decisions made by governments, corporations, markets, and individuals.

AI represents the culmination of a technological progression that began with simple calculators and advanced through programmable software. For decades, computing referred to tools that followed instructions. AI marks a transition to tools that generate internal rules and act on them autonomously. This shift signals the beginning of an era in which decision-making itself becomes automated. The consequences of this change will shape the economic, political, and moral character of the twenty-first century.

AI as the End State of Computing

Artificial Intelligence (AI) began as an academic idea long before it became a commercial force. The term was introduced in 1956 at the Dartmouth Conference by computer scientist John McCarthy, who argued that every aspect of human learning could, in principle, be described with sufficient mathematical precision for a machine to reproduce it. Early experiments reflected this ambition. Simple programs learned to play checkers, solve algebraic equations, and prove basic theorems. Chess soon became a proving ground. A machine could evaluate possibilities at speeds no human could match, demonstrating that computers were capable not only of calculation but of competitive reasoning.

AI developed as the natural continuation of computing rather than a sudden break from it. The first computers functioned as industrial calculators. Software expanded their usefulness by allowing machines to follow instructions automatically. *Machine learning (ML)* took the next step by reducing the need for explicit instructions altogether. Instead of being told exactly what to do, a model could learn patterns from data and apply those patterns to new situations. This transition marks AI as the end state of computing, the point at which software stops behaving as a passive tool and begins functioning as a predictive, adaptive system that learns from experience rather than relying solely on fixed, human-written rules.

Several technological breakthroughs made this shift possible. Cloud computing allowed organizations to process enormous datasets without owning physical infrastructure. Smartphones, social media platforms, and digital services produced an unprecedented volume of behavioral and sensory data. Graphics processing units and specialized AI chips dramatically accelerated neural network training, reducing tasks that once required months of computation to days or even hours. Together, these developments created a global computational backbone capable of supporting AI systems that learn continuously and improve over time.

Traditional AI already forms the backbone of much of today's digital infrastructure. These systems specialize in prediction and classification rather than creativity. They flag fraudulent transactions, interpret medical scans, score credit risk, recommend videos, route delivery fleets, analyze satellite imagery, and optimize supply chains. Their strength lies in statistical precision. By identifying patterns across millions of data points, they perform judgment-heavy tasks at a scale and speed no human workforce could replicate. Much of daily life already depends on these systems, even when their presence goes unnoticed.

A newer branch, commonly called *Generative AI*, expands these capabilities from analysis into creation. Rather than only recognizing patterns, generative models learn the underlying structure of language, images, sound, and code, and then produce new material that aligns with those patterns. Language models generate essays, explanations, and dialogue. Image models produce artwork and design concepts. Music models compose melodies, and code models generate functional software. Generative AI operates by predicting what should come next in a sequence, allowing machines to perform tasks traditionally associated with human creativity and communication.

The shift from Traditional AI to generative systems occurred between 2017 and 2022, following a series of breakthroughs in neural network architecture. In 2017, researchers at Google introduced *the transformer architecture* in a landmark paper titled *Attention Is All You Need.* The transformer enabled models to track context and relationships across long sequences of data in ways earlier neural networks could not. Building on this foundation, OpenAI, an American artificial intelligence research organization, released a series of increasingly capable transformer-based models: GPT-1 in 2018, GPT-2 in 2019, and GPT-3 in 2020. These systems demonstrated that machines could generate coherent language, synthesize ideas, and produce original content.

The public release of ChatGPT in late 2022 marked a turning point. For the first time, advanced generative AI became accessible to ordinary users through a simple conversational interface. A machine could now write extended essays, generate computer code, answer complex questions, and engage in fluent dialogue. What had once been confined to research laboratories entered everyday life. This moment marked the arrival of Generative AI as a distinct technological era rather than a niche research field.

Both Traditional AI and Generative AI are reshaping decision-making across society. Algorithms now monitor financial markets, adjust logistics networks, forecast supply-chain disruptions, and detect security threats in real time. Governments rely on AI to analyze economic conditions, manage public services, and assess risk. Corporations use AI to plan mergers, price products, detect fraud, and automate customer interactions. As more of these activities become automated, AI shifts from assisting human judgment to becoming the primary engine that allocates resources and interprets economic reality.

In this sense, AI completes the long trajectory of computing. The progression began with simple calculation, advanced to programmable instruction, expanded into machine learning, and now culminates in systems capable of adaptive reasoning and creative generation. AI absorbs the functions of all previous computational layers while adding prediction, flexibility, and continuous improvement. It is no longer optional infrastructure. AI now defines how modern organizations operate, compete, and survive.

Machine Learning Fundamentals

Machine learning, often shortened to *ML*, is a branch of *Artificial Intelligence (AI)* that teaches computers to learn patterns from data rather than follow rigid, hand-written instructions. Traditional software depends on explicit rules created by programmers. Machine learning replaces those rules with statistical learning. The system analyzes large numbers of examples, identifies relationships within the data, and builds internal parameters that improve its performance over time. At its core, machine learning is a process of optimization. The algorithm searches for mathematical configurations that minimize error and maximize accuracy.

Modern machine learning systems, including both Traditional AI and Generative AI, learn through three foundational methods: supervised learning, unsupervised learning, and reinforcement learning. Although

Traditional AI and Generative AI differ in what they ultimately produce, they rely on the same learning mechanisms. Both also pass through the same operational stages of training, fine tuning, and inference. What distinguishes Generative AI from Traditional AI is not how learning occurs, but what the system is designed to generate after learning is complete.

In *supervised learning*, the algorithm is trained on labeled data. A system designed to recognize handwritten numbers, for example, is shown thousands or millions of images where the correct number is already known. The model compares its predictions to the correct answers and adjusts its internal parameters until it can reliably identify new images it has never encountered. Many Traditional AI systems rely on supervised learning, including fraud detection tools, medical image classifiers, and credit scoring models.

Unsupervised learning removes labels entirely. Instead of being told what to look for, the system searches for structure on its own. It identifies clusters, similarities, and patterns within the data. A common example is customer segmentation. An unsupervised model can analyze millions of transactions and group customers by behavior without being told in advance what those groups should be. This method is widely used in market analysis, anomaly detection, and exploratory data science.

Reinforcement learning teaches machines through trial and error. The system takes actions within an environment and receives feedback in the form of rewards or penalties. Over time, it learns strategies that maximize long-term success. A well-known example is AlphaGo, which learned to play the game of Go by simulating millions of matches against itself. Through repeated experimentation, the system developed strategies that surpassed those of the world's best human players. Reinforcement learning is also used in robotics, logistics optimization, and autonomous navigation.

Generative AI uses these same learning approaches, particularly *self-supervised learning*, which allows a system to learn structure from raw data without explicit labels. By predicting missing words in text, missing pixels in images, or missing notes in music, the model learns the underlying patterns of language, vision, or sound. This learning process allows generative systems to later create new content that follows the same structural rules as the data they were trained on.

Machine learning systems also move through three operational stages: training, fine tuning, and inference. Training is the initial learning phase in which the model consumes large datasets and adjusts its parameters to reduce errors. Fine tuning adapts this general knowledge to a specific task using more targeted data, such as legal documents, medical records, or financial statements. Inference is the deployment stage, where the trained model applies what it has learned to new inputs in real time.

Traditional AI uses inference to classify or predict outcomes. A fraud detection system evaluates whether a transaction appears suspicious.

A medical model identifies anomalies in imaging scans. A logistics system predicts delivery times based on traffic and weather data. Generative AI uses inference differently. Instead of selecting a category or probability, it predicts what should come next in a sequence. The next element might be a word, a pixel, a musical note, or a line of computer code. In both cases, reliability depends on data quality, data volume, and the computational resources used during training.

These shared foundations produce very different behaviors. Traditional AI systems focus on classification, prediction, and decision support. They determine whether a transaction is fraudulent, which product a user is likely to buy, which route is most efficient, or which video a viewer is likely to watch next. Their strength lies in statistical consistency and speed. They perform judgment-heavy tasks across massive datasets faster than any human analyst could manage.

Generative AI extends machine learning beyond analysis into creation. Rather than only identifying patterns, it learns the structure behind language, images, audio, or code and then produces new material that fits those patterns. Generative systems are not a single technology but a family of models. Some generate images, others generate music, others generate video, and others generate software. *Large Language Models*, often called *LLMs*, are the subset of Generative AI focused specifically on human language.

An LLM is trained on an enormous collections of text so it can learn grammar, meaning, relationships, and contextual patterns. It generates language by predicting the most likely next word in a sequence, but its scale allows it to write essays, summarize documents, answer questions, solve problems, and simulate conversation. LLMs are not separate from machine learning or Generative AI. They are one specialized category within it, optimized for linguistic reasoning.

Machine learning also introduces significant risks. The most fundamental is bias. Every dataset reflects the world from which it was collected, and the world contains incomplete information, uneven representation, and historical inequalities. Opinion polling offers a useful analogy. Polls attempt to measure public sentiment, yet no poll can perfectly represent an entire population. Sampling methods, question framing, and interpretation all introduce distortion. Machine learning systems face the same challenge. If training data reflects biased outcomes or incomplete perspectives, the model will learn and reproduce those patterns, even when no one intends it to do so.

Another major risk is overfitting. *Overfitting* occurs when a model memorizes its training data rather than learning general principles. Such a system performs well on familiar examples but fails in new situations. When deployed at scale, errors caused by bias or overfitting can propagate rapidly through financial systems, medical diagnostics, automated decision-making, or autonomous technologies. Reliable machine learning requires constant evaluation, monitoring, and governance.

The underlying reality is straightforward. Machine learning converts patterns into mathematical rules. Traditional AI applies those rules to analyze and predict outcomes. Generative AI uses them to create new content. Together, these systems form the foundation of automated reasoning in the modern world. They increasingly shape communication, business, governance, and culture, not by understanding meaning, but by statistically modeling it.

The AI Platform Wars

The Artificial Intelligence (AI) landscape is not a single industry or a unified technological project. It is a global contest among powerful platforms, each seeking to build the most capable, scalable, and widely adopted intelligence system. The objective is not merely to release useful tools. It is to establish the foundational models and infrastructure that governments, corporations, and institutions will depend upon as the central decision engines of the modern world.

OpenAI became the most visible participant in this contest by releasing large foundation models that are now used by millions of people each day. These systems generate text, images, software code, and structured reasoning outputs that once required specialized professional training. Their influence flows from three sources: the scale of their training data, the sophistication of their neural architectures, and the ease with which their tools can be accessed. By placing advanced generative capabilities into ordinary workflows, OpenAI demonstrated that general-purpose AI could reshape entire sectors simultaneously rather than advancing in narrow technical niches.

Google DeepMind occupies a different position in the AI ecosystem. From its founding, its primary mission has been scientific discovery rather than immediate commercial deployment. DeepMind first gained global attention with AlphaGo, the system that defeated the world champion in the ancient board game Go. Go had long been considered resistant to machine mastery because the number of possible game states far exceeds what brute-force calculation can evaluate. AlphaGo succeeded by combining deep neural networks with reinforcement learning, allowing it to develop strategies that resembled intuition rather than enumeration.

DeepMind extended this approach with AlphaZero, a system that learned chess, shogi, and Go entirely through self-play, surpassing all human-designed engines without relying on historical game data. Its most consequential breakthrough, however, was AlphaFold. Protein folding had remained an unsolved problem for decades because predicting a protein's three-dimensional structure from its amino-acid sequence is extraordinarily complex. AlphaFold achieved near-experimental accuracy, instantly transforming biology, medicine, and drug discovery. This result demonstrated that AI could generate scientific knowledge rather than merely analyze existing data. DeepMind now applies similar techniques to materials science, weather modeling, fusion energy forecasting, and computational chemistry.

Anthropic represents a philosophical counterbalance within the AI race. Founded by former OpenAI researchers, Anthropic emphasizes alignment, interpretability, and predictable behavior. As AI systems become more capable and autonomous, many institutions require models that can be audited, constrained, and guided by explicit principles. Anthropic introduced the concept of *Constitutional AI*, an approach in which a model's behavior is shaped by written ethical rules rather than by ad hoc human feedback alone. These principles may include fairness, non-manipulation, transparency, and safety boundaries.

Anthropic's models are designed to explain their reasoning, refuse harmful requests, and behave consistently across sensitive domains. The company positions itself not as the largest engineering organization, but as a platform suitable for governments, regulated industries, and institutions that require reliability under legal and ethical constraints. In a climate increasingly concerned with misuse and loss of control, Anthropic sits at the intersection of intelligence and governance.

Meta pursues a fundamentally different strategy centered on

openness and scale. Rather than restricting access to its models, Meta releases powerful systems such as LLaMA, which stands for Large Language Model Meta AI, under open-source licenses. This approach allows researchers, startups, and independent developers to inspect, modify, and deploy advanced language models without dependence on a centralized provider. The result is a global development ecosystem in which Meta's architectures become the default foundation for thousands of downstream applications.

This strategy mirrors earlier open-source successes such as Linux or Android, where widespread adoption itself becomes a competitive advantage. By distributing advanced models broadly, Meta influences research norms, training practices, and technical standards across the industry. Openness accelerates innovation by allowing smaller teams to compete with large corporations using the same underlying technology, even as it raises questions about misuse and governance.

Nvidia provides the computational backbone that powers nearly the entire AI economy. Although it does not build frontier AI models, Nvidia is among the most influential companies in the field. Its graphics processing units, or GPUs, are specialized processors capable of performing thousands of mathematical operations in parallel, making them ideally suited for neural network training.

Nvidia's A100 became the industry standard for large-scale AI development beginning around 2020, powering the first wave of modern large language models and industrial AI systems. Its successor, the H100, introduced in 2022, significantly increased speed and efficiency through improved tensor cores and memory architecture, becoming the backbone of the generative AI expansion that followed. Nvidia's newer GB200 architecture, announced in 2024, combines advanced GPUs with high-performance CPUs into a single module capable of replacing entire racks of older hardware.

Training a frontier artificial intelligence model may require tens of thousands of specialized processors operating continuously for weeks. At this scale, computation becomes a limiting factor rather than a supporting one. Progress is no longer determined only by talent or ideas, but by access to physical infrastructure capable of sustaining extreme computational demand.

Nvidia's influence extends beyond hardware because it has built an integrated technological stack that governs how artificial intelligence is developed, trained, and deployed. At the foundation of this stack is CUDA, short for Compute Unified Device Architecture, introduced in 2006. CUDA allows software developers to write programs that run directly on Nvidia's graphics processors rather than on general-purpose computer chips. This transformation turned GPUs from graphics tools into engines for scientific computing and machine learning.

Once models are trained, TensorRT, released in 2015, optimizes them for deployment. It reduces latency and increases efficiency, allowing AI systems to operate at scale in search engines, recommendation systems, autonomous machines, and financial platforms. Training determines what a model can do. Inference determines whether it can be used economically.

At the system level, NVLink, first deployed in 2014, and InfiniBand, adopted through Nvidia's acquisition in 2019, enable thousands of processors to exchange data at extremely high speeds. These interconnects allow entire data centers to function as unified computational systems rather than collections of isolated machines.

Omniverse, introduced in 2020, extends this ecosystem into simulation. It allows engineers to model factories, robots, logistics networks, and digital twins in shared virtual environments before deploying them in the physical world. This capability links artificial intelligence directly to manufacturing, infrastructure, and industrial planning.

Together, this hardware and software stack determines not only how fast artificial intelligence can advance, but who is capable of advancing it at all. Because nearly every major AI laboratory depends on this ecosystem, Nvidia occupies a position comparable to an industrial bottleneck, controlling the infrastructure upon which modern artificial intelligence depends.

China's AI ecosystem operates in parallel to the Western model and is shaped as much by national strategy as by market competition. Companies such as Baidu, Alibaba, Tencent, and SenseTime develop large models for domestic search engines, e-commerce platforms, logistics systems, facial recognition, and surveillance infrastructure. China's regulatory environment allows access to massive datasets that Western firms cannot legally collect, particularly in areas involving public monitoring, financial behavior, and centralized social platforms.

China treats AI as a strategic asset comparable to nuclear energy or aerospace technology. The state funds chip development, national computing grids, training runs, and specialized research institutes. AI development is tightly integrated with cybersecurity, military planning, supply-chain control, and information governance. A recent participant in the market, DeepSeek, demonstrated that careful optimization and engineering efficiency can allow smaller models and budgets to rival far larger Western systems. Its success suggested that scale alone may not determine future leadership.

Across the global AI ecosystem, companies frequently invest in one another to secure strategic positioning. Each firm seeks to embed its models, hardware, or infrastructure into competitors' workflows. The objective is to establish the foundation layer upon which future applications will be built. Once a dominant foundation model is adopted, entire vertical systems can be constructed on top of it for healthcare, finance, education, defense, logistics, and governance. Control of the foundation layer confers long-term structural power.

The competition follows a familiar economic cycle. Expectations rise faster than practical capability. Valuations inflate. Corrections follow. Weaker firms collapse, while stronger platforms consolidate. What distinguishes the AI cycle from earlier technology booms is the direct involvement of national governments. When states commit billions through procurement programs, defense partnerships, and research funding, AI becomes part of public infrastructure. Firms that secure these relationships gain durability beyond market cycles.

The rapid construction of data centers, research hubs, and computational energy grids reflects a new phase of economic development. The Industrial Revolution built factories, railroads, and steel mills. The digital era builds server farms, fiber networks, and computation clusters. History shows what happens when regions fail to adapt to a new economic foundation. Many industrial towns declined once their core industries lost relevance. Regions that cannot integrate into the computational economy face a similar risk.

The central reality is that AI has become a geopolitical battlefield nested within an economic one. Whoever controls the most capable models shapes communication, scientific discovery, intelligence analysis, and economic planning. Whoever controls the chips determines the pace of progress. Whoever controls the data centers governs the energy and computation upon which societies increasingly depend. The platform wars are not merely about software. They are about the architecture of the next phase of civilization.

AI's Role in Financial Modeling, Markets, and Automation

Artificial intelligence (AI) is now embedded in the core machinery of global finance. It no longer exists at the margins as an experimental aid. It shapes decisions, allocates capital, interprets risk, and automates entire categories of economic activity. In practice, AI has become the analytical nervous system of modern markets.

One of the clearest transformations appears in quantitative trading. AI systems ingest enormous volumes of information in real time, including price movements, order books, macroeconomic indicators, corporate filings, geopolitical events, satellite imagery, and even social media sentiment. These systems detect patterns and correlations far beyond human capacity and execute trades automatically at microsecond speeds. By reacting faster and processing more variables simultaneously, AI-driven trading influences market liquidity, volatility, and price formation across global exchanges.

Credit scoring and lending have undergone a similar shift.

Traditional underwriting relied on limited inputs such as credit history, income verification, and repayment records. AI expands this model dramatically. Modern systems incorporate thousands of variables, including spending behavior, employment stability, transaction patterns, and broader economic conditions. This allows lenders to assess risk more precisely and extend credit to borrowers previously excluded from formal financial systems. At the same time, this expansion introduces new risks. As decision-making migrates into automated systems, outcomes become harder to inspect, surveillance grows more pervasive, and individuals increasingly struggle to contest judgments rendered by algorithms whose logic they cannot see, audit, or meaningfully challenge.

In consumer and institutional investing, automated advisory systems, often called *robo-portfolios*, rely on AI to recommend asset allocations based on individual goals, timelines, and risk tolerance. These systems rebalance portfolios continuously, manage tax efficiency, and respond dynamically to market conditions. Services once available only to wealthy clients through high-fee advisors are now accessible to millions at low cost. As a result, a significant portion of public investment behavior is increasingly shaped by a small number of algorithmic frameworks.

Macroeconomic forecasting has also shifted toward AI-driven analysis. Central banks, multinational corporations, and sovereign wealth funds use AI models to interpret inflation trends, recession indicators, supply-chain disruptions, and geopolitical risk. These models synthesize information from commodity markets, shipping data, satellite imagery, weather systems, and global news flows. The advantage is speed and scope. The risk is that national policy decisions may increasingly reflect algorithmic interpretation rather than human judgment grounded in historical and political context.

Within financial institutions, AI has transformed back-office operations that once required entire departments. Reconciliation, transaction monitoring, compliance screening, fraud detection, and regulatory reporting are now handled by systems that operate continuously at machine speed. Anti–money laundering models evaluate thousands of transactions per second, flagging anomalies for review. Insurance underwriting and claims processing are moving toward full automation. While these changes reduce cost and increase efficiency, they also eliminate traditional middle-class career pathways that once anchored financial employment.

AI has become equally essential to fraud prevention and cybersecurity. Modern anomaly-detection systems learn what constitutes normal behavior and immediately identify deviations. They detect fraudulent payments, uncover identity theft, and flag cyber intrusions long before humans notice irregularities. Digital banking and payment platforms now depend on these systems to maintain trust and operational stability.

Corporate finance has also been reshaped. AI models assist executives in evaluating mergers and acquisitions, restructuring supply chains, adjusting pricing strategies, and monitoring competitive threats. Strategic planning increasingly relies on algorithmic forecasts and scenario modeling. Decisions that once rested primarily on human analysis are now informed, and in some cases driven by machine-generated insight.

The cumulative effect is structural. AI functions as the brain of the financial system. It interprets global conditions, directs capital flows, and performs judgments that once required armies of analysts, traders, and economists. Firms that integrate AI operate faster, manage risk more efficiently, and outcompete those that do not. Markets themselves behave differently because a large share of activity is driven by automated decision-making rather than human discretion.

This transformation carries ethical and social consequences. When algorithms govern access to credit, investment opportunity, and risk classification, accountability becomes fragmented. Bias embedded in training data can scale instantly. Automated trading can amplify volatility or trigger cascading failures. As AI replaces human judgment, entire professional classes face displacement. The system grows more efficient, but also more brittle, because errors propagate at digital speed.

The central reality is this. Financial intelligence has become computational. Markets, institutions, and governments increasingly rely on automated reasoning to organize economic life. In the *Techonomic* era, the power to compute aligns closely with the power to rule, because whoever controls the algorithms controls the allocation of wealth, opportunity, and risk.

Chapter V

Fintech Frontiers

Banking Reinvented

For most of modern history, banking was defined by physical architecture. Marble columns, brass doors, and polished stone interiors were not merely decorative, they communicated stability, permanence, and trust. A bank was a physical place anchored to geography. Customers stood in line, filled out paper forms, waited for approvals, and interacted face to face with staff. The building itself functioned as a guarantor of safety. When smartphones placed a bank branch in every pocket, that centuries-old relationship began to dissolve. Banking shifted from physical locations to digital experiences, and the industry has spent the last decade adapting to that change.

The first visible transformation was the collapse of the branch network. Between 2010 and 2025, thousands of bank branches across the United States closed as customer activity migrated to mobile devices. This transition did more than improve convenience, it reshaped expectations. Checks could be deposited by camera. Identity could be verified through an app rather than across a desk. Transfers that once required days began to settle instantly. Customer service moved from physical counters and call centers to automated chat systems. The smartphone did not support the branch model. It replaced it.

As digital banking became standard, banks began to recognize that their real advantage was not real estate, but infrastructure. This realization led to API-driven banking. An *Application Programming Interface (API)* is a software bridge that allows one system to request services from another. Instead of a customer entering a branch and completing paperwork, a mobile application can now ask a bank's systems to verify identity, check balances, initiate payments, or evaluate credit in real time through these digital gateways.

To understand APIs, imagine a bank's internal systems divided into many small service windows, each responsible for a single function. One window verifies identity. Another confirms available funds. Another processes a payment. Another updates transaction records. Each window is an API endpoint. When a user taps "Send $50," the application may call several APIs in rapid sequence: one to authenticate the user, one to confirm the balance, one to execute the transfer, and one to update the ledger. To the user, the action appears instantaneous. Beneath the surface, multiple systems coordinate in milliseconds.

APIs transformed banking functions into modular software components that approved partners could integrate directly into their products. Banks began operating more like cloud platforms than storefronts. JPMorgan Chase leveraged its scale to build a dominant mobile banking ecosystem. At the same time, infrastructure firms such as Plaid and Stripe created universal connectors that allowed startups to access financial rails without negotiating separately with every bank.

This shift enabled a new generation of fintech platforms. Cash App combined payments, banking, investing, and social interaction into a single interface. PayPal expanded from online checkout into a global financial network offering lending, credit products, merchant tools, and cross-border transfers. Robinhood placed commission-free trading on millions of phones. These platforms performed many of the same functions as banks while avoiding the traditional label. They issued cards, held balances, verified identity, processed loans, and moved money entirely through software.

Initially, traditional banks viewed fintech firms as competitors. Over time, the relationship evolved into partnership. Banks needed fintech innovation and user experience. Fintech firms needed bank charters, regulatory compliance, and access to the Federal Reserve system. A hybrid model emerged. Banks provided the regulated foundation. Fintech platforms supplied the customer-facing layer. This framework, commonly called *banking-as-a-service*, now supports neobanks, gig-worker payroll systems, construction-lending platforms, and hundreds of specialized financial applications.

This evolution reflects a deeper structural change. Banking is no longer defined primarily by people managing assets. It is defined by algorithms managing flows of information. When a customer opens a mobile banking app, they engage with a decision engine that verifies identity, analyzes risk, calculates balances, detects fraud, and routes transactions in milliseconds. Paperwork has been replaced by machine logic.

Fintech also exposes the limits of the existing financial system. It modernizes banking, but it still depends on licensed institutions, regulators, and corporate ownership. Companies such as PayPal, Square, SoFi, Robinhood, and Coinbase extend the reach of finance without altering its underlying structure.

Decentralized finance (DeFi) takes a different approach. Rather than building companies on top of banks, DeFi replaces the company with code. Smart contracts on blockchains such as Ethereum run lending markets, exchanges, stablecoin systems, and collateralized vaults without executives or employees. Protocols like Aave, Uniswap, Curve, and MakerDAO operate through transparent algorithms and community voting rather than corporate boards. There are no branches, compliance departments, or

CEOs. There is only software.

Fintech upgrades the financial system that already exists. DeFi attempts to construct a parallel system that operates independently of banks and governments.

Ultimately, fintech broke the monopoly of geography. A branch no longer defines a bank's reach. A well-designed application does. The winners are not only the institutions with the most real estate, but also those with the fastest transfers, the most seamless interfaces, and the most powerful APIs. Banking has largely ceased to be a place you visit. It has become a service you carry, and increasingly a system that operates whether you notice it or not.

Embedded Finance in Everyday Life

One of the most consequential changes in modern finance has been the shift from finance as a destination to finance as an invisible layer embedded within daily activity. For most of the twentieth century, people encountered financial services through explicit institutions such as banks, branches, and card networks. Financial action required deliberate engagement. A person went to a bank, filled out forms, signed documents, and waited for approval. Today, most people interact with financial systems dozens or even hundreds of times a day without consciously engaging with a bank at all.

Every ride taken through Uber, every food order placed on DoorDash, every purchase completed on Amazon, every subscription renewed on Spotify, and every item bought through a social media platform triggers a sequence of financial processes. Payments are authorized, fraud checks are performed, accounts are credited or debited, and records are updated. These actions occur silently in the background. This is *embedded finance*, the integration of financial services directly into the software environments people already use.

The concept behind embedded finance is straightforward. Instead of requiring users to leave an application to complete a financial task, the financial action occurs inside the experience itself. When a person exits an Uber, payment is settled automatically. When an order is placed on Amazon, payment authorization and fraud detection occur instantly without additional steps. Many platforms go further. A coffee shop application may allow users to preload balances, earn rewards, or trigger automatic payment when an order is picked up. Retail platforms embed installment options directly at checkout. Delivery platforms offer instant access to earnings. In each case, finance is present, but it is no longer visible.

This transformation was made possible by companies that built the financial plumbing of the internet. Stripe allows applications to accept payments without building their own payment infrastructure. Plaid enables secure connections between applications and users' bank accounts. Marqeta powers customizable debit cards for platforms such as Square and Instacart. These firms do not replace banks. They abstract them. Complex banking functions are converted into standardized software components that developers can integrate quickly. What once required years of negotiation, regulatory approval, and infrastructure investment can now be implemented in weeks through a small number of APIs.

As a result, non-financial companies have become financial operators without formally becoming banks. Uber delivers near-instant payouts to drivers. Shopify offers merchant loans and cash-advance products. Apple provides installment payments and credit services through Apple Pay Later. Amazon extends credit lines to sellers and embeds financing into its marketplace. These companies did not originate as financial institutions, yet they now move and manage enormous financial flows. Finance is no longer their product. It is a tool that strengthens their core business models.

For consumers, embedded finance has permanently reshaped expectations. Delays feel unacceptable. Manual steps feel outdated. Users expect payments to settle instantly, refunds to process automatically, and financing to appear at the precise moment it is relevant. In this environment, the most successful platforms are not necessarily banks, but companies that deliver the smoothest financial experiences while remaining largely invisible.

The deeper implication is that finance has become contextual. Individuals no longer go to a bank to access financial services. Financial services appear automatically when needed. Financing emerges at checkout. Insurance appears during travel booking. Payments occur during transportation. Earnings settle immediately after work is completed. Banking is no longer a separate activity. It has become a background utility, as constant and unnoticed as electricity.

Embedded finance fulfills fintech's original promise. Financial services are faster, simpler, and woven into everyday life. Consumers do not want to perform banking tasks. They want banking to happen without interruption. The institution fades from view. The function remains.

Digital Wallets and Neobanking

The idea of a wallet was once literal. It was a physical object used to carry cash, cards, and identification. Over the last decade, that concept has shifted from leather and plastic to software. Digital wallets now live on phones and browsers, and they function as far more than payment tools. They have become gateways into financial life, authentication systems, and, increasingly, containers for digital identity. Platforms such as Apple Pay, Google Wallet, Cash App, Venmo, and Alipay are reshaping how people store value, move money, verify identity, and interact with the financial system.

The rise of *digital wallets* began when smartphones incorporated secure hardware components capable of storing encrypted credentials. This allowed a phone to perform the same authentication role as a physical card, often with greater security. A fingerprint, facial scan, or device-based token could now authorize a transaction. This development enabled contactless payments, peer-to-peer transfers, and mobile checkout at scale. Over time, wallets expanded beyond payments. They began to store transit passes, event tickets, loyalty programs, health credentials, and government-issued identification. The digital wallet evolved into a portable identity layer, gradually replacing the collection of cards people once carried in their pockets.

As wallets matured, many transformed into financial super-apps. Cash App allows users to send money, receive paychecks, invest, file taxes, and purchase Bitcoin (BTC) within a single interface. Alipay integrates bill payments, commerce, microloans, insurance, travel booking, and savings products into one ecosystem. In parts of China, Southeast Asia, and Africa, the digital wallet functions as the primary banking interface. Many users in these regions never interacted with a traditional bank branch. Their entire financial lives exist on a phone.

Crypto wallets introduced a different and more radical concept: self-custody. Unlike wallets tied to banks or payment networks, crypto wallets hold digital assets directly through cryptographic private keys. Ownership is proven mathematically rather than institutionally. *Seed phrases* replaced passwords. Users became responsible for safeguarding their own assets. These wallets support cryptocurrencies, tokens, NFTs, and on-chain credentials. They represent an early form of decentralized digital identity, where access and ownership depend on cryptography rather than corporate or government systems.

Alongside wallets, a new type of institution emerged: the neobank. Platforms such as Chime, Revolut, Monzo, and N26 operate entirely through mobile applications without physical branches. By avoiding real estate and legacy infrastructure, neobanks offer faster onboarding, lower fees, and real-time account insights. While deposits are typically held by partner banks with formal charters, neobanks differentiate themselves through interface design, transparency, and automation. Features such as instant balance updates, spending analytics, automatic savings tools, and real-time alerts became standard expectations rather than premium services.

The growth of digital wallets and neobanks has raised new regulatory questions. When a platform offers checking accounts, credit tools, and payment services but does not hold a banking charter, how should it be supervised? When companies such as Apple, PayPal, or Cash App manage customer balances and process billions in transactions, should they face the same oversight as traditional banks? Regulators across the world are still adapting to entities that behave like banks without fitting historical definitions.

The broader shift is clear. Trust has moved away from physical institutions and toward digital interfaces. The brand that mediates a person's financial life may no longer be a century-old bank with marble columns. It may be a mobile application downloaded in minutes. Financial identity has become software, and the new gatekeepers are the companies that deliver the most reliable, intuitive, and frictionless experiences, not the ones that own the most buildings.

Lending Platforms and Robo-Advisors

For most of modern banking history, borrowing money required a personal encounter. A prospective borrower sat across from a loan officer, completed paperwork, submitted documents, and waited while a human evaluated creditworthiness. This process began to change in the late 2000s with the rise of *peer-to-peer lending*, commonly known as *P2P lending*. Platforms such as Prosper and LendingClub allowed individuals to lend directly to other individuals through online marketplaces rather than through banks.

Under this model, a borrower seeking a loan might be matched with dozens or hundreds of small lenders, each contributing a portion of the total amount. A five-thousand-dollar loan could be assembled from many smaller commitments. The platform handled identity verification, payments, and servicing, while interest flowed directly to the lenders. This approach removed the bank as an intermediary and reframed lending as a digital marketplace rather than a branch-based service.

P2P lending laid the groundwork for a deeper transformation: algorithmic underwriting. Traditional credit evaluation relied heavily on narrow indicators such as credit scores, payment history, and reported income. Fintech platforms expanded this lens dramatically. Modern underwriting systems analyze alternative data, including cash-flow patterns, frequency of income deposits, bill-payment consistency, spending behavior, employment volatility, and account usage over time. This method, often called *alternative data modeling*, produces a more granular picture of financial behavior than legacy credit scoring alone.

In many cases, this expansion improved access to credit. Younger borrowers, gig-economy workers, recent immigrants, and others with limited credit histories could now be evaluated based on observable financial activity rather than the absence of a traditional score. At the same time, this shift raised concerns about surveillance, opacity, and fairness. When credit decisions are made by algorithms evaluating hundreds or thousands of variables, borrowers may not understand why they were approved or denied, or how to challenge an outcome.

Investing evolved alongside lending. A new category of platform, known as the *robo-advisor*, emerged through services such as Betterment,

Wealthfront, and Schwab Intelligent Portfolios. These systems replaced human financial advisers with automated portfolio management. After answering a series of questions about goals, time horizon, and risk tolerance, users received diversified investment portfolios built and maintained by algorithms.

Robo-advisors automated tasks that once justified high advisory fees. They rebalanced portfolios as markets moved, optimized for tax efficiency through tax-loss harvesting, and adjusted allocations over time as user circumstances changed. What had once been available primarily to wealthy clients became accessible to middle-income households and first-time investors. Investing shifted from periodic decision-making to continuous algorithmic management.

Together, these developments mark a structural change in how credit and investment guidance are delivered. Decisions that were once made by loan officers and financial advisers are now made by software systems operating at scale. Access to borrowing and wealth-building tools increasingly depends on how algorithms interpret behavior rather than how humans evaluate character or circumstance.

This transformation has expanded inclusion and lowered costs, but it has also centralized decision-making within opaque systems that few users fully understand. The gatekeepers of modern lending and investing are no longer individuals sitting behind desks. They are predictive models embedded in platforms, quietly shaping who receives capital, under what conditions, and at what cost.

New Customer Banking

One of the most consequential shifts in consumer finance over the past decade did not originate inside traditional banks. It emerged at the checkout screen. *Buy Now Pay Later* services, commonly abbreviated as *BNPL*, introduced a new model of short-term consumer credit that felt simpler, more transparent, and less intimidating than credit cards. Companies such as Klarna, Afterpay, and Affirm allowed shoppers to divide purchases into fixed installments, often with no interest if payments were made on time.

For a generation shaped by student debt, rising living costs, and skepticism toward revolving credit, BNPL felt intuitive. Instead of borrowing against an open-ended credit line, consumers saw a concrete schedule. A purchase might be presented as four payments of twenty-five dollars rather than a single one-hundred-dollar charge. The obligation felt bounded and predictable. There was no abstract interest rate compounding in the background, only a visible sequence of payments tied to a specific purchase.

Merchants adopted BNPL rapidly because it changed buyer behavior. When cost is reframed as smaller installments, hesitation decreases. Conversion rates rise. Average order sizes increase. Checkout itself becomes a moment of automated credit underwriting. Algorithms evaluate a customer's profile in seconds, drawing on prior repayment behavior, transaction history, and platform-specific data. Approval or rejection occurs instantly, without human review. What once required a credit application now happens invisibly inside the purchase flow.

This frictionless design reshaped consumer expectations. Credit was no longer something applied for in advance. It appeared contextually, precisely when a consumer was deciding whether to buy. In this environment, borrowing became embedded in consumption rather than separated from it. The line between payment and credit blurred.

BNPL also altered financial behavior in less visible ways. Because installment plans were easy to open and felt modest individually, many users began managing multiple BNPL obligations simultaneously across different platforms. A consumer might remain current on each plan while still accumulating a complex web of short-term liabilities. Unlike credit cards, which aggregate balances into a single account, BNPL obligations are fragmented and dispersed, making total exposure harder for consumers to track.

Regulators began to take notice. Concerns emerged around late fees, missed payments, inconsistent reporting, and consumer understanding. Some BNPL providers initially avoided reporting repayment behavior to credit bureaus, which meant borrowers could accumulate obligations without those debts appearing in traditional credit profiles. Others began experimenting with partial reporting or internal scoring systems that functioned independently of legacy credit infrastructure.

In response, BNPL firms increasingly repositioned themselves as data-driven lenders rather than simple payment tools. They expanded underwriting models, integrated repayment history into risk assessments, and explored partnerships with credit agencies. In doing so, they quietly reshaped how creditworthiness is defined for millions of younger consumers who may never rely primarily on credit cards.

This development signals a broader transformation in consumer banking. Credit decisions are no longer anchored to long-term relationships with financial institutions. They are contextual, instantaneous, and algorithmic. The modern consumer encounters credit not through a bank officer or account manager, but through a checkout interface governed by software. Banking, once organized around accounts and branches, is reorganizing around moments, behavior, and data-driven judgment.

Chapter VI

Algorithmic Power

Social Media Propaganda and Bot Farms

Social media transformed propaganda from a slow, labor-intensive practice into an automated system capable of operating at global scale. In earlier eras, influence required physical infrastructure. Newspapers had to be printed. Radio stations needed transmitters. Television required studios, licenses, and distribution agreements. Propaganda moved at the speed of human labor and institutional coordination. Today it moves at the speed of software.

Modern platforms rank and distribute content through algorithms optimized for engagement rather than accuracy or civic value. What rises to prominence is not what is most truthful, but what generates the strongest reaction. This single design choice converts influence into a computational problem. The question is no longer how to persuade millions of people individually, but how to trigger the algorithmic systems that decide what millions of people will see.

At the center of this environment are *bot farms*. A *bot* is a software program designed to imitate human activity online. It can post content, like or share messages, follow accounts, reply to comments, and engage in conversations. A single operator can control thousands of such accounts simultaneously. When deployed in coordinated waves, these bots manufacture the appearance of popularity, urgency, or consensus. An idea appears to be everywhere not because people support it, but because software is repeating it at scale.

Real-world examples illustrate how this operates. The Russian Internet Research Agency used coordinated bot networks during the 2016 and 2020 United States elections to amplify divisive content and deepen political polarization. Similar commercial bot farms operate in India, China, Eastern Europe, and parts of Southeast Asia, selling political engagement, follower counts, and artificial popularity to clients ranging from political campaigns to corporations. Bots do not tire, hesitate, or require payment in proportion to their output. They can flood online spaces continuously, overwhelming genuine discussion through sheer volume.

More advanced campaigns rely on *Coordinated Inauthentic Behavior* (CIB). These operations blend automated accounts with real human operators to create networks that appear organic. Accounts post at realistic intervals, maintain personal histories, and engage in casual conversation alongside political messaging. Facebook, now Meta, has publicly identified CIB networks linked to foreign intelligence agencies, private influence firms, domestic political groups, and corporate interests. These networks do not merely spread messages. They manufacture agreement, simulate grassroots support, and suppress opposing viewpoints by drowning them in coordinated responses.

Foreign governments exploit these techniques with increasing precision. China uses coordinated activity across platforms such as TikTok, WeChat, and X to promote pro-state narratives and deflect criticism of the Communist Party. Russia deploys influence operations designed to fracture trust in Western institutions and exacerbate cultural divisions. Iran and North Korea employ similar strategies to undermine confidence in foreign governments and media. These campaigns require no physical presence. A single well-timed post, amplified by algorithmic systems, can reach millions within hours.

Domestic governments are not absent from this landscape. Modern states increasingly shape internal narratives through strategic messaging, information framing, and platform partnerships rather than overt censorship. In the United States and other modernized nations, this has taken the form of coordinated public-information campaigns, selective amplification of official narratives, and collaboration with social media platforms to flag or suppress content deemed harmful, misleading, or destabilizing. During periods of crisis such as public health emergencies, elections, or national security events, information flows are actively managed to maintain social order, public compliance, and institutional trust. The methods differ from foreign propaganda in intent and transparency, but the mechanisms are often the same: algorithmic visibility, repetition, and authority signaling.

Algorithms accelerate the effects of both foreign and domestic influence. Recommendation systems guide users toward content that produces strong emotional engagement, regardless of origin. A person who watches a single political video may quickly be directed toward increasingly narrow or extreme material. This pattern has been documented across platforms such as YouTube, TikTok, and Facebook. Groups seeking influence study these mechanisms closely. They craft messages, memes, and videos designed not to persuade through argument, but to exploit algorithmic incentives. The result is a gradual funnel in which beliefs shift incrementally while the user experiences the process as self-discovery rather than manipulation.

Underlying this entire system is the *economics of attention.* Platforms profit by keeping users engaged for as long as possible. Content that provokes outrage, fear, tribal loyalty, or excitement spreads faster because it stimulates the nervous system. Propagandists have learned to reverse-engineer these incentives. Messages are optimized for algorithmic visibility rather than truth. A sensational falsehood can outcompete accurate reporting simply because it is more emotionally stimulating.

The consequence is a new information battlefield. Propaganda no longer depends on printing presses, broadcast towers, or large editorial staffs. It depends on algorithms, data, and computational force. Influence can be deployed instantly, continuously, and at scale. Control over attention has become a strategic asset, and public discourse has become a contested digital terrain governed less by human deliberation than by machine learning systems optimized for engagement.

The AI-Generated World

Artificial Intelligence (AI) has elevated digital influence to a level that no human workforce could sustain. In earlier eras, propaganda required writers, editors, designers, videographers, and coordinated distribution teams. Each message carried a real cost in time and labor. AI removes that constraint. It can generate text, images, audio, video, and entire online conversations at scale, allowing a single operator to produce the output of an entire media organization in minutes.

This shift matters because AI-generated content often appears convincingly human. Language models can write persuasive essays, respond emotionally, and adapt tone to different audiences. Image models can fabricate realistic photographs or political imagery. Voice models can clone speech patterns from short samples. Video models can animate faces and gestures. Together, these tools erase many of the visual and linguistic cues people once relied on to judge authenticity.

One of the most powerful applications is personalized political persuasion. AI systems analyze online behavior, search history, purchasing patterns, and social connections to build psychological profiles. Instead of broadcasting one message to millions of voters, AI generates millions of tailored messages designed to exploit specific emotional triggers. A person motivated by fear receives a different message than someone motivated by economic anxiety or national pride. Persuasion becomes individualized rather than collectivized.

This approach was visible in early form during the Cambridge Analytica scandal, which emerged publicly in 2018. Cambridge Analytica harvested personal data from nearly eighty-seven million Facebook users through a personality quiz presented as academic research. That data was used to construct psychological profiles and deliver targeted political advertisements designed to influence behavior rather than inform judgment. While this occurred before modern Generative AI, it demonstrated how data-driven persuasion could bypass traditional political messaging and operate beneath public awareness.

AI has since made this process faster, cheaper, and more precise. Generative systems can now produce unlimited variations of a message, test engagement responses in real time, and refine outputs automatically. What once required analysts and strategists now happens continuously through machine learning feedback loops.

AI also enables synthetic engagement. Likes, comments, shares, followers, and trending signals can all be generated automatically. Content that appears to have thousands of responses may be supported by only a small number of operators. Once engagement crosses a threshold, recommendation systems may promote the content organically, amplifying it further. Genuine users often interpret high engagement as social proof, unaware that the popularity itself was manufactured.

More advanced systems fabricate entire online communities. AI-driven accounts can debate topics, share personal stories, disagree politely, and reinforce group norms. A forum may appear to reflect grassroots discussion when every participant is algorithmic. This makes it increasingly difficult to distinguish authentic public sentiment from coordinated synthetic movements created by political actors, corporations, or foreign governments.

Moderation struggles to keep pace. Human reviewers cannot evaluate millions of posts generated within minutes. Automated filters often fail when AI content mimics natural language patterns closely. Because recommendation algorithms reward emotional intensity, sensational AI-generated material often spreads faster than corrective information. The system learns from engagement and becomes more persuasive over time.

The most significant consequence is the emergence of an artificial public sphere. For the first time, a substantial share of online speech may not originate from human beings at all. Public opinion becomes difficult to measure when voices are algorithmically produced. Movements can be simulated rather than built. Cultural trends can be engineered rather than emerging organically.

Influence now depends less on ideology or truth and more on mastery of algorithms and computational resources. Digital reality is no longer merely a reflection of human behavior. It is increasingly shaped, filtered, and directed by machines. As AI expands deeper into communication systems, the boundary between genuine discourse and automated persuasion becomes harder to perceive, marking a fundamental shift in how power operates in the information age.

Shaping Public Opinion

Algorithms do more than display information. They actively shape what people see, what they consider important, and what they eventually believe. Every click, pause, search, like, and share becomes a data signal. These signals train the system to predict what will hold a user's attention next. Over time, this creates a feedback loop. The user teaches the algorithm, and the algorithm teaches the user. The influence feels natural because it unfolds gradually, guided by mathematical optimization rather than overt persuasion.

Personalization is the core mechanism behind this process. Social networks, search engines, video platforms, and news aggregators tailor content to individual users. This personalization increases relevance and convenience, but it also fragments shared reality. A person who watches political videos is shown more political content. A person who engages with conspiratorial material is guided toward increasingly similar material. The system does not evaluate truth or balance. It evaluates engagement. As a result, two people living in the same home can occupy entirely different informational environments.

Predictive modeling intensifies this effect. Machine learning systems estimate how an individual will respond emotionally or behaviorally to certain messages. Political campaigns, corporations, and advocacy groups use these predictions to adjust what content is delivered to whom. An undecided voter may receive reassuring or optimistic messages, while a committed partisan may receive content designed to heighten outrage or reinforce loyalty. These interventions are subtle. Most users do not recognize that they are being guided toward particular emotional states rather than informed about competing viewpoints.

Algorithmic Power

Platform design further amplifies this influence through dopamine-driven feedback. Algorithms prioritize content that produces strong emotional reactions because emotional responses keep users engaged. Anger, fear, humor, tribal identity, and moral outrage consistently outperform calm analysis or nuance. This reward structure reshapes public discourse. Extreme voices gain visibility. Moderate perspectives struggle to compete. Creators who provoke reaction are promoted, while those who encourage reflection are often ignored.

Algorithms have also become the new cultural gatekeepers. In earlier eras, editors, producers, and publishers decided what reached the public. Today, automated ranking systems determine which articles appear at the top of search results, which videos are recommended, and which posts surface in feeds. These decisions are made continuously, based on predicted engagement rather than editorial judgment. This grants platform owners immense cultural power. Visibility itself becomes a form of authority.

The influence extends beyond opinion into identity. Many people form their beliefs, communities, and sense of belonging through online spaces. When algorithms shape what they see and who they interact with, they also shape perceptions of what is normal, popular, or acceptable. Subcultures are reinforced. Polarization accelerates. Worldviews solidify through repetition rather than deliberation.

Machine learning does not merely reflect public opinion. It participates in its formation. By deciding what is amplified and what fades into obscurity, algorithms influence elections, consumer behavior, cultural trends, and social norms. This power no longer belongs exclusively to governments or traditional media institutions. It resides within automated systems that operate continuously, invisibly, and at planetary scale.

Deepfakes & Synthetic Reality

Deepfakes are artificial intelligence generated videos, images, and audio that make it appear as though a person said or did something that never occurred. These fabrications rely on advanced neural networks, often using *Generative Adversarial Networks (GANs),* to replicate facial expressions, voice patterns, and physical movement with increasing realism. What began as an experimental novelty has become a powerful instrument for misinformation, fraud, and political manipulation. A convincing deepfake of a public figure can circulate globally before journalists or fact-checkers are able to respond, creating confusion and damaging trust in authentic media.

Voice cloning expands this threat further. With only a brief audio sample, AI systems can reproduce a person's voice and generate entire conversations or speeches that sound genuine. Criminal groups have already used cloned executive voices to authorize fraudulent wire transfers.

Security agencies warn that similar techniques could be used to issue fake emergency alerts, military orders, or diplomatic statements. Because people instinctively trust familiar voices, synthetic audio undermines one of the most basic signals of authenticity.

Another emerging category is the creation of digital humans and AI generated influencers. These entities are virtual personalities built entirely in software. They promote products, interact with followers, and comment on cultural events. Companies favor them because they are controllable, consistent, and immune to scandal. They never age, never violate contracts, and never deviate from approved messaging. While often presented as entertainment or marketing innovation, they blur the line between genuine communication and algorithmically engineered persuasion.

Synthetic media also presents challenges for law enforcement and legal systems. Courts have long treated photographs, recordings, and video as reliable evidence. Deepfakes erode this foundation. A fabricated video could be used to discredit a witness or falsely implicate an innocent person. Investigators must now treat digital evidence with skepticism that was unnecessary in previous eras. Journalists face similar difficulties as verification becomes more complex in an environment where convincing fabrications can be produced quickly and cheaply.

The political consequences are equally severe. Manipulated videos and audio can inflame tensions, influence elections, incite violence, or destroy reputations. Coordinated groups can release synthetic content engineered to provoke emotional reactions that spread faster than corrections can follow. When citizens cannot trust what they see or hear, confidence in public institutions weakens. People begin to assume that any damaging evidence may be fake, even when it is real.

The deeper danger is the erosion of a shared reality. Societies that have a democratic process depend on a basic agreement about facts. Synthetic media destabilizes that agreement. If convincing forgeries are common and denials are always plausible, truth itself becomes contested terrain. Preserving trust will require new verification systems, widespread digital literacy, and trusted methods of authentication capable of distinguishing human reality from machine fabrication.

Deepfakes and synthetic reality illustrate the cultural consequences of advanced artificial intelligence. They demonstrate how technology can manipulate perception, distort communication, and undermine trust at scale. As these systems improve, the challenge will not only be detecting falsehoods but preserving the conditions under which truth can still be recognized and defended.

The Ethics of Algorithmic Influence

Algorithms no longer merely recommend content or organize information. They shape beliefs, attitudes, and perceptions at scale. Human attention itself has become contested terrain. Governments seek to influence what their populations see in the name of stability, security, or national interest. Technology companies seek to influence behavior to maximize engagement, revenue, and platform growth. The ethical crisis emerges from this overlap. Influence is no longer exercised openly. It is embedded in code, optimized invisibly, and deployed continuously.

This conflict is intensified by the imbalance of power between nation states and the technocratic class that designs and controls the dominant platforms. Governments attempt to regulate digital influence through law, oversight, and enforcement, yet algorithms evolve faster than legislation can respond. Meanwhile, technology firms operate across borders, shaping the beliefs of billions without direct democratic accountability. The question is not whether influence exists, but who ultimately exercises it. Are elected institutions governing information flows, or are private systems quietly shaping political outcomes by directing public attention?

Transparency complicates the issue further. Algorithms are complex, adaptive, and often proprietary. Meaningful inspection is difficult even for their creators. Excessive transparency risks exposing systems to manipulation by adversaries. Insufficient transparency leaves societies vulnerable to unseen persuasion. This tension creates space for abuse on multiple fronts. Governments may exploit opaque systems to suppress dissent. Corporations may exploit them to maximize profit regardless of social cost. Citizens remain largely unaware of how deeply these mechanisms guide their perceptions.

Privacy and data collection deepen the ethical stakes. Algorithmic systems build detailed psychological profiles using behavioral data collected across platforms, devices, and environments. This data can be used to predict emotional responses, influence decisions, and shape identity over time. When governments access such data, surveillance becomes granular and persistent. When corporations control it, persuasion becomes personalized and relentless. In both cases, the individual becomes the object of optimization rather than the subject of choice.

Some argue that stronger regulation is necessary to protect societies from manipulation by both state and corporate actors. Others warn that granting governments extensive control over digital systems risks entrenching political power and silencing dissent. The ethical dilemma has no easy resolution. Power concentrated in private platforms threatens democratic oversight. Power concentrated in the state threatens individual liberty.

The unresolved question is which force will dominate the coming era. Will nation states succeed in asserting authority over algorithmic systems, or will technocratic networks continue to shape public consciousness beyond the reach of law? More fundamentally, will individuals exist within these systems as sovereign actors at all, with meaningful control over their data, their algorithmic environments, and clear knowledge of who is influencing them and how they are being observed? How societies answer these questions will determine whether digital systems serve human agency or quietly erode it. The ethics of algorithmic influence are not abstract. They define the boundary between persuasion and control in the *Techonomic* age.

Chapter VII

Data Supremacy

Why Data Outranks Capital and Labor

In the twenty-first century, data has become the most valuable strategic resource on earth. Nations once competed for oil fields, trade routes, and industrial capacity. Today they compete for something less visible but far more influential: the continuous streams of behavioral information generated by billions of people each day. Every action that passes through a digital device becomes a signal that can be stored, analyzed, and converted into prediction. This predictive capability now underpins modern economic and political power.

The value of data lies not only in its scale but in its precision. Modern systems collect what is often described as *behavioral telemetry*, the high-resolution digital traces that reveal how people move, react, search, scroll, and decide. These signals include geolocation trails, device fingerprints, dwell time, swipe patterns, hesitation intervals, and micro-interactions that expose preferences with remarkable accuracy. A single data point reveals little. Millions of such signals, combined over time, produce behavioral models that no human organization could construct manually.

Data also differs from traditional resources because it scales without physical constraint. Land is finite. Minerals must be extracted. Industrial output requires factories, labor, and logistics. Data, by contrast, can be duplicated and transmitted across borders instantly. When a platform gains users in Brazil, India, or Nigeria, its predictive systems improve immediately, without constructing a warehouse or shipping a product. This global scalability is unique to digital resources and explains why dominant platforms expand so rapidly once network effects take hold.

What makes data strategically decisive is its ability to reduce uncertainty. Every institution operates with incomplete information. In earlier eras, governments and corporations relied on surveys, limited samples, and human intuition. Today they draw on real-time behavioral telemetry generated by millions of individuals. Companies forecast demand, test products before launch, detect risk earlier, and influence consumer behavior with mathematical confidence. Governments apply similar methods to monitor economic conditions, anticipate unrest, and assess social sentiment.

The advantage compounds through a force often described as *data gravity*. The more data a platform collects, the more valuable its systems become. Increased value attracts more users. More users generate more data. Over time, this feedback loop creates a gravitational center around platforms with the deepest datasets and most advanced analytics, pulling developers, advertisers, and institutions into their orbit.

In this environment, power is no longer measured solely by armies, factories, or gross domestic product. It is measured by the richness of data streams and the sophistication of the systems that interpret them. Institutions that possess continuous, diverse, and high-quality data gain visibility into society that competitors cannot match. They can anticipate behavior, shape decisions, and influence markets in ways that widen the gap between data-rich and data-poor actors.

Data has become the central resource of the *Techonomic* era. It transforms uncertainty into prediction, prediction into influence, and influence into power. Those who collect, store, and analyze the most information now exert disproportionate control over markets, culture, and geopolitical outcomes.

Global Data Infrastructure

Data supremacy is not achieved through collection or analysis alone. It depends on the physical infrastructure that stores, moves, and processes information across the globe. Behind every cloud platform, recommendation engine, and artificial intelligence system stands an industrial network of data centers, fiber-optic lines, and undersea cables. These structures form the digital backbone of the modern economy, even though most people remain unaware of their existence.

The most critical component of this infrastructure is the modern *data center*. These facilities resemble industrial warehouses, filled with servers, cooling systems, and high-voltage electrical equipment that operate continuously. Increasingly, companies rely on *colocation centers* rather than building private server farms. In these arrangements, firms rent secure racks, power, and cooling within shared facilities. This model allows startups and mid-sized companies to operate with the same physical reliability as global corporations, while concentrating control over data storage in the hands of a small number of infrastructure providers.

Data centers alone are insufficient without the networks that connect them. The majority of international digital traffic travels through undersea *fiber-optic cables*, thousands of miles long, laid across the ocean floor. These cables carry more than ninety-five percent of global data flows. On land, they connect to *terrestrial fiber networks*, including stretches of unused or reserved capacity known as *dark fiber*. Corporations lease dark fiber to create private, high-speed routes that move information faster than public networks allow, reducing latency and strengthening real-time analytical capabilities.

Fiber-optic cables are the physical arteries of the digital world. Unlike copper wires, which transmit electrical signals, fiber transmits data as pulses of light through strands of ultra-pure glass roughly the width of a human hair. Light-based transmission offers vastly higher bandwidth and lower signal loss, enabling enormous volumes of information to move with minimal delay.

Terrestrial fiber networks crisscross continents, often running alongside highways, rail corridors, and utility rights-of-way. Many of these fiber bundles contain far more strands than are actively used. The unused strands are referred to as dark fiber because they are not "lit" by transmission equipment. Corporations, cloud providers, universities, and government agencies lease this dark fiber to establish dedicated data corridors that bypass congestion and grant direct control over information flow. These private routes support applications ranging from high-frequency trading to real-time artificial intelligence inference at industrial scale.

Most people are unaware that the utility poles outside their homes carry multiple layers of infrastructure. The highest lines belong to electric utilities and carry high-voltage power. Beneath them sit telecommunications lines, including copper telephone cables, coaxial internet lines, and bundles of fiber-optic strands. Dark fiber occupies this middle layer. It is separate from the electrical grid but shares the same physical pathways, quietly forming part of the national data backbone.

These neighborhood fiber bundles often contain dozens or even hundreds of strands, only a fraction of which are activated. The remaining fibers remain dark until leased by organizations requiring private, high-speed connectivity. From these local networks, long-haul terrestrial fiber routes extend across states and regions, eventually connecting directly to undersea cable systems that link entire continents. Together, these networks form the physical foundation of the global internet.

Undersea fiber-optic cables represent the most critical and least visible component of global data infrastructure. The United States connects to Europe through systems such as MAREA, built by Microsoft and Meta and completed in 2017, which runs from Virginia to Spain. The Dunant cable, funded by Google and completed in 2020, links Virginia to France. The Grace Hopper cable, completed in 2022, connects New York to the United Kingdom and Spain.

Across the Pacific, major routes include the FASTER cable, completed in 2016 and funded by Google, NEC, and Japanese telecom firms, which links Oregon to Japan and Taiwan. The JUPITER cable, completed in 2021 and backed by NTT, SoftBank, Amazon, and Meta, connects California to Japan and the Philippines. Additional systems extend from Florida and Virginia into South America, including AMX-1, completed in 2014, and BRUSA, completed in 2018, which connect the United States directly to Brazil, Colombia, and Mexico. These cables carry cloud services, financial transactions, artificial intelligence training data, video streaming, military communications, and the daily digital activity of billions of people.

Taken together, these networks form the hidden physical foundation of data supremacy. Ownership, construction, and influence over these routes confer geopolitical leverage. Controlling the pathways of global data flow determines speed, security, and sovereignty across entire economies. The competition over undersea networks is therefore not merely about bandwidth. It is about national power and control over the information pipelines that sustain modern civilization.

Because of their importance, nations now treat digital infrastructure as strategic territory. Governments compete to host cable landing stations, the points where undersea cables connect to terrestrial networks. Control over these stations resembles control over ports and shipping lanes in earlier eras. Similarly, countries seek to attract data centers, which bring investment, energy demand, and geopolitical influence. These assets define which nations function as global data hubs.

American cloud firms such as Amazon, Google, and Microsoft are constructing data campuses the size of industrial parks, complete with dedicated substations, advanced cooling systems, security perimeters, and private fiber connections. Comparable projects are expanding across Europe, India, and Southeast Asia. At the same time, China's Belt and Road Initiative includes a Digital Silk Road that installs undersea cables, cloud infrastructure, and telecommunications systems across dozens of countries.

This physical infrastructure is the skeleton of the digital world. Every search query, financial transaction, artificial intelligence inference, and social media interaction passes through these systems. Without them, algorithmic platforms and prediction engines would fail instantly. Data supremacy is therefore not only about building powerful models. It is about constructing and controlling the factories, pipelines, and power grids that keep those models alive.

The race to build and govern this infrastructure reveals a fundamental truth of the *Techonomic* era. Digital power is physical power. Data centers, fiber networks, and undersea cables are the industrial projects of the twenty-first century, shaping global influence as decisively as railroads and oil pipelines shaped the twentieth.

Monetizing Data

Once data is collected and transported, it becomes a financial asset. The modern digital economy rests on a process often described as data valorization, the conversion of raw behavioral signals into revenue-generating products and services. Companies no longer profit simply by storing information. They profit by extracting patterns, generating predictions, and selling those predictions to institutions that depend on behavioral forecasting.

The largest digital platforms operate vast attention markets in which individual moments of focus are bought and sold in real time. When a user scrolls through social media or performs a search, automated bidding systems activate behind the scenes. Advertisers compete for the opportunity to display a message to that specific user at that specific moment. This process, known as *real-time bidding*, is possible only because platforms track detailed behavioral telemetry. Attention becomes a tradeable commodity. Systems such as Google Ads and Meta Ads function less like traditional advertising agencies and more like automated exchanges for human behavior.

Beyond advertising, a secondary industry has emerged around the direct buying and selling of personal information. Data brokers such as Acxiom, Experian, and Equifax assemble extensive digital profiles using thousands of data sources. These firms combine shopping histories, credit records, public filings, geolocation trails, and online interactions into composite dossiers. The resulting profiles are sold to marketers, insurers, lenders, political consultants, and financial firms. Much of this ecosystem relies on what is often called data exhaust, information generated incidentally while users engage in unrelated activities. This exhaust is collected, refined, and monetized without most individuals ever realizing it occurred.

Retail companies have also become data firms in practice. Loyalty programs offered by Walmart, Target, and major grocery chains are designed not only to reward repeat purchases, but to capture detailed information about consumer behavior. Each scanned barcode generates a signal that feeds predictive pricing systems, inventory optimization models, and personalized marketing engines. Routine shopping activity becomes part of a behavioral refinery that extracts insight from everyday decisions.

Financial institutions participate through *spending telemetry* and transaction analytics. By analyzing purchase histories, banks infer customer risk, predict life events such as job changes or relocations, and offer targeted financial products. Mobile apps, streaming services, and online platforms often sell metadata such as device characteristics, session duration, or approximate location to advertising networks and commercial data exchanges. In many cases, a single digital action passes through a chain of intermediaries, each extracting value before the user becomes aware any information was transferred.

Artificial intelligence has accelerated this process by enabling the creation of inference products, predictions derived from data rather than the data itself. A retailer may purchase a probability score estimating whether a household is likely to buy furniture within the next thirty days. A lender may acquire a repayment-risk model rather than raw transaction histories. Political consultants may buy behavioral clusters built from similarity patterns rather than traditional demographic categories. These inference products allow firms to monetize future behavior, turning prediction itself into a standalone commodity.

Most individuals never share directly in the value they generate. Their behavioral exhaust is captured and monetized by platforms that treat personal data as an extractable resource. In response, a counter-movement has begun to emerge around self-sovereign identity and decentralized data markets. These approaches argue that individuals should be able to own, license, and profit from their personal information rather than surrender it by default. If such systems mature, they could reverse decades of asymmetry and establish the first genuine markets where users retain control over the value derived from their own behavior.

In the *Techonomic* era, data monetization functions as the engine of the digital economy. It transforms human experience into financial instruments, converts attention into auctionable inventory, and turns prediction into profit. As long as data can be captured and analyzed at scale, it will continue to generate economic value that rivals, and in many cases exceeds, that of traditional industries.

Sovereign Data Systems

As data becomes the central strategic resource of the twenty-first century, nation-states are no longer willing to leave its control entirely in private or foreign hands. Governments increasingly view data as a matter of national security, economic sovereignty, and political stability. In response, many countries are constructing sovereign data systems, national frameworks designed to govern how information is collected, stored, processed, and accessed within their borders. This marks a departure from the early internet era, which favored open flows and private infrastructure, toward a world where states assert authority over digital space.

One of the most common tools is *data localization.* These policies require companies to store and process certain categories of data within the country where it is generated. India's Reserve Bank mandates local storage of financial transaction data. Indonesia enforces domestic handling of payment information. The European Union, through the General Data Protection Regulation (GDPR), restricts the transfer of personal data outside the bloc unless strict safeguards are met. These rules give governments legal leverage over digital activity and prevent foreign firms from extracting value without domestic oversight.

Another pillar of sovereign data strategy is the development of *national digital identity systems.* These centralized platforms authenticate citizens for banking, healthcare, taxation, voting, and public services. India's Aadhaar system, which contains biometric profiles for over one billion people, is the largest example in existence. Estonia's e-ID program allows citizens to sign contracts, access government services, and participate in civic life entirely online. These systems function as internal data engines, granting governments detailed visibility into economic activity and social behavior while increasing citizen dependence on state-managed digital infrastructure.

Some countries pursue sovereignty by building state-aligned cloud infrastructure. Rather than relying exclusively on American hyperscalers such as Amazon, Google, or Microsoft, governments seek domestic or regional alternatives. France and Germany launched Gaia-X as an attempt to establish a European-controlled cloud and data-sharing ecosystem governed by local standards. China integrates its major technology firms into a unified national cloud strategy, ensuring that sensitive data remains under state jurisdiction. These sovereign clouds serve not only as storage platforms but as enforcement mechanisms, allowing governments to regulate access, monitor data flows, and exclude foreign competitors.

Sovereign ambition also extends into artificial intelligence. Nations increasingly treat large datasets as strategic inputs for AI development. Governments assemble national training datasets drawn from healthcare records, transportation systems, *agricultural telemetry*, and demographic databases. These datasets support domestic AI industries and reduce dependence on foreign models. China uses state-collected data to train industrial automation systems and surveillance platforms. The United Kingdom aggregates national health data to support medical AI research. Control over data enables control over models, and control over models translates into economic and military leverage.

These developments reshape geopolitics. Digital borders now matter alongside physical ones. States negotiate data-sharing agreements, restrict foreign platforms, and build firewalls that regulate the movement of information. The global digital economy is fragmenting into spheres of influence: an American cloud-centric sphere, a European regulatory sphere, and a Chinese state-engineered sphere. Each reflects a distinct philosophy of how data should be governed, monetized, and deployed.

In this environment, data is no longer merely a commercial asset. It is a geopolitical resource. Nations that build coherent sovereign data systems gain leverage over global information flows, influence the development of artificial intelligence, and shape the rules of the digital economy. The contest of the coming decades will not be defined solely by who collects the most data, but by who commands the infrastructure and authority that determine how data is used.

Privacy vs. Progress

The rise of the tech economy has created a persistent tension between innovation and individual rights. Digital systems offer convenience, personalization, and automation, but these benefits depend on continuous access to personal information. Navigation apps require location data. Recommendation systems depend on behavioral history. Smart assistants rely on voice, usage patterns, and contextual cues. In this environment, privacy is no longer a peripheral legal issue. It is a structural question about how modern digital systems function.

Governments have attempted to address these concerns through privacy regulation. Europe's General Data Protection Regulation (GDPR), California's Consumer Privacy Act (CCPA), and China's Personal Information Protection Law (PIPL) were designed to give users greater control over how their data is collected, processed, and shared. These frameworks mandate transparency, consent mechanisms, and security standards. In practice, however, regulation often lags behind technological change. Platforms adapt faster than laws can be updated, creating enforcement gaps where personal data continues to circulate through complex corporate ecosystems.

This has produced what critics describe as *privacy theater*. Companies comply with formal requirements while maintaining data-intensive business models beneath the surface. Consent banners appear, privacy policies expand, and opt-in dialogs proliferate, yet the underlying extraction of behavioral data continues largely uninterrupted. The appearance of control improves, but the balance of power remains unchanged.

Interface design further weakens meaningful consent. Default settings frequently encourage data sharing, and many permission flows rely on dark patterns that steer users toward accepting broad access. Most people approve these requests without fully understanding their scope or consequences. The legal framework of consent remains intact, but practical agency shifts toward the platform.

Privacy also involves economic tradeoffs. Restricting data collection reduces the effectiveness of AI-driven services. A navigation system without location tracking cannot optimize routes. A recommendation engine without behavioral history cannot personalize content. Limiting data can enhance individual protection, but it also constrains functionality. Societies are forced to balance efficiency, innovation, and privacy rather than maximizing all three simultaneously.

In response, new approaches to digital autonomy are emerging. *End-to-end encryption*, *decentralized identity systems*, and s*elf-sovereign data wallets* aim to return control to users. These technologies challenge centralized data models by allowing individuals to verify identity, share credentials selectively, and retain ownership over personal information. If widely adopted, they could alter how data value is distributed across the digital economy.

Yet technical solutions alone cannot resolve the deeper tension. The dominant economic incentives of the tech economy remain aligned toward continuous behavioral analysis and algorithmic governance. Even when privacy rights expand, platforms still profit from prediction, personalization, and influence. Without structural changes to these incentive systems, privacy protections risk becoming administrative layers that soften public concern without altering core business practices.

The central question is whether technological progress can coexist with meaningful privacy. Can societies preserve innovation without granting institutions unchecked authority over personal data, or will continuous monitoring become a permanent condition of digital life. In the *Techonomic* era, the issue is no longer only access to information. It is whether individuals retain power over how their behavior is interpreted, predicted, and used.

Chapter VIII

Convergence

The New Digital Superstructure

The modern world is no longer built primarily on factories, shipping ports, or industrial corridors. It is built on computation. Behind every application, algorithm, financial product, and digital interaction stands a vast technological scaffold that did not exist a generation ago. This superstructure forms the environment in which the *Techonomic* era takes shape. It is an interconnected system of data centers, undersea cables, identity frameworks, payment rails, cloud platforms, artificial intelligence systems, and public blockchains that now operate as the foundational machinery of economic and social life.

Most people encounter only the surface of this system. A mobile application loads instantly. A recommendation appears in a feed. A payment settles without delay. A digital wallet verifies identity with minimal friction. These experiences are possible only because deeper layers of infrastructure coordinate millions of operations every second. Data centers function as industrial factories for computation, absorbing electricity and converting it into continuous streams of digital processing. Undersea fiber-optic cables act as the circulatory system of the global economy, carrying nearly all international data traffic between continents. Cloud platforms provide the computational environment where companies deploy AI models, operate financial networks, store behavioral telemetry, and manage the digital identities of billions of users.

This infrastructure does not operate as a collection of isolated components. It behaves as a single, integrated system. Smartphones, sensors, and connected devices collect raw behavioral signals. Fiber networks transport those signals to data centers. Machine learning systems transform them into prediction and classification. Blockchain networks record transactions and anchor trust. Financial APIs convert prediction into economic action by issuing payments, evaluating credit, or allocating capital in milliseconds. Each layer feeds the next. As digital activity expands, the underlying systems become more powerful and more tightly coupled.

In earlier eras, industrial strength was measured by physical output. In the *Techonomic* era, strength is measured by the density and sophistication of digital infrastructure. A country with advanced cloud regions, high-capacity fiber routes, and large-scale AI training clusters possesses advantages comparable to the industrial monopolies of the past. Companies with deep data pipelines, optimized compute environments, and integrated financial rails operate at a scale that smaller competitors cannot match. Digital infrastructure is no longer a supporting function. It has become the primary engine of national competitiveness and corporate strategy.

This superstructure shapes how modern life operates. It determines the speed of communication, the efficiency of markets, the organization of financial systems, and the capacity of governments to function in a fast-moving digital environment. It is the hidden architecture beneath the visible surfaces of everyday experience. As more of the world moves online, this superstructure increasingly governs the global economy. It defines the conditions under which the century ahead will unfold.

The Techonomic Loop

The *Techonomic Loop* is a conceptual model for understanding how the digital age operates as an integrated system rather than as a collection of separate technologies. Its emergence was not accidental, nor was it geographically neutral. The system first consolidated in the West, and particularly in the United States, because America exited the twentieth century as the dominant military, economic, and technological power. Europe, once a peer competitor, was materially and demographically devastated by the Second World War in the 1940s and spent decades rebuilding under security and financial dependence. The United States' remaining systemic rival, the Soviet Union, collapsed in the early 1990s before the digital era fully matured. As a result, the foundational architectures of the modern digital economy were established under American governance, institutional norms, and strategic priorities. Other nations entered the digital age not as co-authors, but as adapters, imitators, or resisters of a system whose rules had already been written.

The loop begins with governance. Governments fund infrastructure, establish standards, enforce regulations, and define national priorities. These decisions shape the computational layer that powers the digital economy. Computation generates and processes data, which feeds artificial intelligence systems capable of interpreting human behavior at scale. AI systems then shape the attention environment, where recommendation engines, search algorithms, and ranking systems influence what people see and how they act. This attention layer feeds directly into platforms and financial systems, where commerce, fintech, neobanking, blockchain networks, and decentralized finance convert digital behavior into economic activity. The outcomes of these systems return to governments and corporations, informing new policies, investments, and strategic decisions. The loop functions as a continuous cycle, not a vertical hierarchy, with each layer reinforcing the next.

At the foundation of the loop sits the governance layer, the political and strategic framework that enables the entire system. No digital ecosystem emerges spontaneously. National security priorities, industrial policy, defense spending, and regulatory design determine which technologies are developed, which infrastructures are built, and which firms rise to global prominence. The governance decisions of the twentieth century shaped the semiconductor supply chain, the early internet, cloud computing architectures, payment networks, and privacy regimes that structure digital life today. Governance initiates the loop and later absorbs its effects, reshaping each successive cycle.

From governance emerges the computational layer, the physical machinery of digital civilization. Data centers, graphics processing clusters, cloud platforms, semiconductor fabrication plants, undersea cables, terrestrial fiber networks, mobile infrastructure, and satellite systems form the industrial base that sustains all digital activity. This layer provides the raw capacity required for every higher function. Without computation, there can be no artificial intelligence, no digital finance, no platforms, and no scalable digital economy. As computational capacity increases, each subsequent layer expands in capability.

Computation produces and processes the signals that populate the data layer, where behavioral telemetry becomes the raw material for prediction. Smartphones, applications, platforms, payment systems, and sensors generate continuous data streams that record how billions of people move, search, purchase, communicate, and decide. Machine learning systems transform these streams into classification, forecasting, recognition, and generative output. As computational power grows, AI models become more capable. As AI capabilities improve, they incentivize the collection of even more data. The loop tightens, accelerating with each iteration.

These predictive systems feed into the attention layer, which mediates between algorithmic logic and human perception. Recommendation engines, search rankings, social feeds, and content-generation systems determine which information becomes visible and which fades into obscurity. This layer does not merely present options. It shapes perception, preference, and behavior. It functions as the tuning mechanism of the loop, amplifying certain signals while suppressing others.

The outputs of the attention environment give rise to the platform layer, where economic activity is organized. Technology firms, fintech applications, neobanks, digital wallets, blockchain networks, and decentralized finance protocols operate within the behavioral patterns shaped by attention systems. Payments, lending decisions, trading activity, identity verification, and incentive structures emerge within this environment. Finance becomes software, embedded directly into daily experience. Whether interacting with Apple Pay, Amazon, Cash App, Shopify, Coinbase, or a decentralized protocol, individuals participate in systems that translate behavior into economic outcomes in real time.

These outcomes complete the loop by flowing back into governance. Tax revenues, market concentration, data accumulation, technological dependency, cultural influence, and national competitiveness return to states and institutions. Governments respond by updating regulations, funding new infrastructure, supporting research initiatives, and redefining strategic priorities. What begins as governance returns as governance, reshaped by the very systems it enabled.

The *Techonomic Loop* is therefore not a chain of command but a self-reinforcing circuit. Computation, data, artificial intelligence, attention, platforms, finance, and governance evolve together. Each layer depends on the others. Each accelerates the next. Together they form the structural logic of the modern digital world, explaining why technologies once treated as separate domains now converge into a single, integrated system.

Autonomous Markets and Algorithmic Governance

As the layers of the *Techonomic Loop* reinforce one another, they give rise to a new economic environment often described as *cybernetic markets.* These are markets that operate through continuous feedback loops between computation, data, artificial intelligence, and protocols rather than through human intermediaries alone. They are not theoretical. They emerge directly from the integration of global computation, behavioral telemetry, predictive models, programmable contracts, and platform infrastructure into a single operational system. The result is an economy in which decisions once made by people are executed automatically by systems tuned to real-time signals.

At the core of this environment are *automated decision systems* (ADS). These systems begin with algorithmic interpretation. AI models analyze patterns across millions of transactions and behavioral traces to generate credit evaluations, pricing adjustments, fraud alerts, routing decisions, and forecasts with machine-level precision. Underwriting becomes instantaneous. Loan approvals no longer depend on a banker's discretion. E-commerce pricing adjusts dynamically in response to user behavior. Insurance premiums update according to fresh telemetry rather than annual reviews. Markets respond continuously because the infrastructure observes continuously.

This automation reshapes the structure of commerce itself. Payments settle through digital gateways embedded directly inside applications. Programmable contracts execute trades automatically when predefined conditions are met. Identity is authenticated through digital credentials rather than physical signatures or in-person verification. Firms rely on automated decision systems to plan inventory, allocate capital, structure supply chains, and assess labor needs. Human oversight remains present, but the operational core runs at machine tempo, driven by constant computation.

Public blockchains introduce a parallel form of autonomy by

shifting trust from institutions to code. Smart contracts enforce agreements without brokers or clearinghouses. Decentralized exchanges allow assets to trade continuously without centralized market makers. Tokenized systems distribute incentives automatically, adjusting fees or rewards according to rules embedded directly into protocols. These systems do not replace traditional finance. They extend it, offering a programmable and transparent method of economic coordination that operates alongside established institutions.

Governments remain central actors within this environment through what is increasingly described as *algorithmic governance.* Rather than directing markets manually, states shape the conditions under which automated systems operate. National identity frameworks become gateways for economic participation. Regulations define how data may be used, how automated systems can make decisions, and how platforms must authenticate and report activity. Sovereign cloud infrastructures, national data strategies, and public AI initiatives establish boundary conditions for autonomous economic behavior. Governance shifts from supervising individual industries to regulating the information flows that drive them.

At the same time, platforms that mediate daily life begin to resemble civic infrastructure. Search engines determine which knowledge becomes visible. Social networks structure public discourse. Payment platforms function as financial utilities. These organizations enforce rules, define norms, and mediate access at a scale once associated with governments. They are not states, but their infrastructures perform governance-like functions. Their algorithms determine what is promoted, what is restricted, and what is monetized. They allocate attention, access, and economic opportunity across billions of users.

As a result, the boundary between markets and institutions becomes increasingly porous. Markets operate through algorithms. Platforms mediate social and economic life. Governments govern through data, identity, and infrastructure as much as through law. This convergence is not the outcome of a single innovation. It is the cumulative result of integrating computation, data, AI, protocols, and platforms into a unified operating environment.

Cybernetic markets, automated decision systems, and algorithmic governance are therefore not speculative futures. They describe the present structure of economic coordination in a world where digital systems interpret behavior, allocate resources, and enforce rules at planetary scale. They are the functional expression of convergence itself.

Life Inside the Converged System

For most people, the transformation of the digital world is not experienced as infrastructure or protocol. It is experienced as convenience. The mechanics of convergence remain largely invisible while daily life is shaped through what scholars describe as *technological mediation*, the process by which systems interpret behavior, guide choices, and automate tasks that once required effort or expertise. The boundaries between work, commerce, communication, and finance blur because these activities now operate within a single, continuous digital environment.

Payments illustrate this shift clearly. A person buys groceries, books a ride, orders a meal, or sends money to a friend without entering a bank or interacting with a physical terminal. Identity verification, fraud analysis, balance checks, and settlement occur beneath the surface through application programming interfaces and predictive models. The user encounters none of this machinery. They tap a screen, the transaction completes, and life continues uninterrupted.

Communication follows the same pattern. Messaging platforms integrate text, video, audio, and artificial intelligence assistance into unified interfaces. Translation and transcription occur automatically. Photo libraries organize themselves. Social feeds adjust continuously to individual habits and interests. The line between information and interpretation grows thin as algorithms personalize content in real time. The digital environment adapts to the user rather than requiring the user to navigate complexity.

The workplace reflects this convergence as well. Cloud systems synchronize files across devices. Collaboration platforms allow teams to operate across continents as if they occupy a single room. Artificial intelligence systems draft documents, summarize meetings, and automate routine functions. Tasks that once required physical proximity now unfold inside computational frameworks that treat geography as largely irrelevant.

Daily movement through the physical world is also guided by unseen digital systems. Navigation applications interpret traffic conditions and reroute automatically. Transit systems authenticate passengers through mobile credentials. Retail environments link inventory systems to personalized recommendations. Smart devices coordinate lighting, temperature, and security inside homes. Each interaction generates data that feeds back into the systems that manage it, creating cycles of continuous adjustment and refinement.

Finance is woven into these routines as well. Paychecks arrive in digital wallets. Investments rebalance through automated portfolios aligned with long-term goals. Micro-savings tools sweep small amounts into diversified holdings. Rental applications, insurance quotes, and credit approvals rely on predictive models that evaluate behavioral patterns rather than static records. Economic decisions that once demanded deliberate planning now occur through systems designed to anticipate needs.

Identity is emerging as the next major layer of convergence. Digital wallets increasingly store credentials, licenses, and permissions alongside payment methods. As governments and platforms integrate identity systems more deeply, participation in digital life begins to resemble movement through a continuous authentication layer. Access to services, communication, travel, and finance increasingly depends on persistent digital verification.

Life inside the converged system feels effortless, but that effortlessness rests on dense technological coordination operating just below awareness. Human behavior is translated into digital signals and met with personalized responses. The world becomes easier to navigate because the environment adapts continuously to each individual. Through the convergence of artificial intelligence, data, finance, identity, and infrastructure, the digital realm ceases to be a separate domain. It becomes the organizing principle of everyday life.

The Architecture Still Being Built

The convergence of artificial intelligence, data infrastructure, digital finance, and global platforms may appear complete, but the system remains unfinished. The digital superstructure that supports modern life continues to expand, reorganize, and adapt to rising demands for scale, efficiency, and coordination. New layers emerge as older ones evolve. The architecture of the *Techonomic* era is not a fixed design. It is a living system shaped by technology, economics, and governance acting upon one another.

Digital identity remains one of the least settled components of this structure. Governments and platforms are building frameworks that authenticate individuals across services, institutions, and borders. These systems will determine how people access healthcare, banking, education, employment, and communication. They will also influence where authority ultimately resides, whether with states, corporations, or hybrid arrangements that blend the two. The future of digital identity will shape how individuals relate to the converged system and how power is exercised within it.

Programmable finance is also still in formation. Smart contracts, tokenized assets, stablecoins, and decentralized networks demonstrate how financial logic can operate through software rather than institutions. Yet these systems are increasingly intertwined with banks, regulators, and global payment networks. The final configuration of a programmable monetary ecosystem has not yet emerged. Its direction will depend on how these tools integrate with real economic activity and how existing institutions respond to digital financial architecture.

Artificial intelligence continues to move beyond large, general-purpose models toward specialized systems embedded directly into transportation, scientific research, manufacturing, logistics, and public administration. As these systems integrate more deeply into physical and institutional environments, AI becomes a structural component of society rather than a collection of discrete tools. Its long-term trajectory is not only greater capability but deeper entanglement with the infrastructure that governs daily life.

Global digital infrastructure is expanding at the same time. Data centers grow larger and more numerous. Undersea cables extend to new regions. National cloud strategies mature. Edge computing pushes processing power closer to vehicles, factories, sensors, and users. The physical footprint of the digital world continues to multiply, reshaping which regions emerge as technological hubs and which fall outside the primary flows of global data.

A central force in this evolving architecture is *interoperability*. The ability of identity systems, financial networks, AI platforms, and national data frameworks to interact smoothly will determine how effectively the digital world functions. Institutions that cannot interoperate risk fragmenting the global system. Those that can will define standards for communication, authentication, intelligence, and value exchange. The nations and platforms that set these standards will exert influence over the global economy for decades.

The result is a world in which the foundations of everyday life are being rebuilt in real time. Technologies once treated as separate domains are becoming components of a unified system that organizes communication, commerce, mobility, identity, and knowledge. The convergence described in this chapter is not an endpoint. It marks the beginning of a long transformation in which digital infrastructure becomes the dominant environment of human society.

The architecture of the *Techonomic* era is still forming. The decisions made now will shape how this structure supports innovation, opportunity, and the flow of information. With each passing year, the system becomes more interconnected, drawing more aspects of life into its networks and extending digital influence over the physical world. Convergence is the moment these forces begin to operate as a unified whole. The task ahead is understanding how this architecture will guide the century to come.

Chapter IX

Rules & Risks

Governance Challenges in a Borderless Tech World

The central question facing modern governance is straightforward: what fails when national law collides with borderless technology? Traditional rulemaking rests on *regulatory territoriality*, the assumption that laws apply within physical borders and that institutions exercise authority over people, firms, and systems located inside them. This framework evolved for factories, shipping lanes, and broadcast networks. It becomes unstable when applied to technologies that operate everywhere at once.

Artificial intelligence models may be trained in one country, deployed in another, and influence users across the entire world simultaneously. Crypto networks operate through thousands of distributed nodes with no physical headquarters. Consumer platforms serve hundreds of millions of users at once, subject to dozens of legal systems that were never designed to function together. Digital activity no longer maps cleanly onto geography, yet law remains territorial by design.

Once platforms become borderless, accountability becomes unclear. When personal data crosses jurisdictions, which privacy regime governs its use? When a crypto exchange incorporated in a permissive jurisdiction serves users in Europe or the United States, which regulator holds authority? Enforcement becomes uneven. Firms respond by relocating to friendlier environments, a behavior known as *regulatory arbitrage*. Some nations lower standards to attract innovation. Others attempt to extend their laws beyond their borders. The result is a fragmented global rulebook applied to systems that function as unified networks.

This tension intensifies around questions of sovereignty. Should platforms themselves bear primary responsibility because they control the infrastructure? Should national governments assert authority even when their legal power does not align with how digital systems operate? Or should supranational bodies such as the European Union or the United Nations establish shared frameworks for technologies that cannot be confined to any one state? Each approach has emerged in practice, yet none resolves the problem on its own.

New forms of liability follow. Which jurisdiction prosecutes when an AI system causes harm across multiple countries? Who is responsible when a decentralized protocol triggers financial losses for users distributed across continents? The traditional legal concept of jurisdiction no longer aligns with where digital actions occur or where their consequences are felt.

The underlying reality is simple. Governance systems built for a world of borders now confront technologies that ignore borders entirely. Unless regulatory institutions evolve at least as quickly as the systems they oversee, the gap between innovation and accountability will continue to widen. That gap carries direct consequences for sovereignty, civil rights, economic stability, and public trust in the digital age.

Systematic Risks in AI and Crypto Compliance

The next governance question is equally direct: what fails when compliance systems collide with rapid technological change? Artificial intelligence and crypto no longer exist outside regulatory attention. In the United States, the Securities and Exchange Commission (SEC) treats certain digital assets as securities, while the Commodity Futures Trading Commission (CFTC) classifies others as commodities. In Europe, the Markets in Crypto-Assets (MiCA) framework establishes a unified rulebook for token issuance, stablecoins, custody, and disclosure. At the international level, the Financial Action Task Force (FATF), formed by the Group of Seven (G7: United States; Canada, United Kingdom; France; Germany; Italy; Japan), pressures governments to apply strict anti–money laundering and identity verification standards across borders.

Despite this activity, the sector remains shaped by what economists describe as regulatory overhang. *Regulatory overhang* refers to rules that are anticipated, partially defined, or unevenly enforced. Firms operate under the expectation that future interpretations may differ sharply from present guidance. This uncertainty introduces risk even when no laws have formally changed.

Artificial intelligence developers face a similar environment. The European Union's Artificial Intelligence Act introduces risk classifications, documentation requirements, and transparency obligations. The United States relies primarily on nonbinding guidance and agency-specific standards. China emphasizes content supervision, model alignment, and state oversight. Companies building global products must navigate incompatible regimes while attempting to anticipate rules that have not yet fully emerged.

This uncertainty reshapes competition. Large firms absorb the cost of compliance by hiring legal teams, auditors, and risk officers. Smaller firms often cannot. Startups delay launches, restrict features, or avoid entire jurisdictions because compliance costs exceed their operational capacity. Regulation becomes not only a safeguard but a market filter that determines which actors survive.

A clear illustration appeared in 2023, when the SEC expanded its interpretation of what constitutes a security. Billions of dollars in digital asset valuations disappeared within days. The underlying technologies did not change. The code did not change. The risk emerged from a shift in legal framing rather than from technical failure. Innovation was not the destabilizing force. Regulatory uncertainty was.

In an ecosystem defined by speed, ambiguity becomes structural. When compliance rules remain unsettled, uncertainty itself becomes a cost. Credibility depends not only on technical performance but on regulatory resilience. The challenge for firms is no longer simply following existing rules. It is operating in an environment where the rules themselves are still forming, and where retrospective enforcement can redefine past behavior overnight.

Cybersecurity Threats and Vulnerable Infrastructure

The next question is practical rather than theoretical: what fails when digital infrastructure is compromised? As identity systems, financial networks, and operational controls migrate online, cybersecurity becomes the constitutional layer of trust in the *Techonomic* era. Yet every new connection expands what security specialists call the *attack surface*, the total set of points through which a system can be accessed or exploited. This phenomenon, known as attack surface expansion, is unavoidable in a world built on cloud platforms, automated protocols, and billions of connected devices.

Interconnection magnifies consequence. A failure in one component can propagate rapidly across systems that depend on it. *Ransomware* attacks now shut down hospitals, schools, and energy infrastructure. *Phishing schemes* and *credential theft* drain digital wallets and compromise personal accounts. Smart contract vulnerabilities can erase millions of dollars within minutes once a flaw is discovered. Data breaches expose biometric identifiers or medical records that cannot be reset or reissued. In such environments, attackers do not need to compromise the most important system. They need only exploit the weakest connected link.

The 2021 Colonial Pipeline incident illustrates this dynamic. A single compromised password enabled a ransomware attack that disrupted fuel distribution across the Eastern United States. The attackers never accessed the operational control systems that manage pipeline flow. They penetrated the broader information environment surrounding it. A peripheral failure produced central disruption. Panic buying began within hours. Fuel shortages spread across multiple states. Federal authorities intervened to stabilize supply. The event demonstrated how minor digital vulnerabilities can trigger large physical consequences when infrastructure is tightly coupled to information systems.

These realities force a shift in defensive strategy. Security can no longer rely on perimeter defenses alone. It must be embedded at every layer. *Zero trust architecture* assumes that internal systems may already be compromised and therefore verifies every action continuously. Smart contracts require formal verification before deployment to eliminate logical flaws that attackers can exploit. Artificial intelligence systems must protect model weights, training data, and inference pipelines against tampering or extraction. Cybersecurity is no longer a technical add-on. It is systemic risk management for an economy that operates at digital speed.

Ethical Dilemmas in Techonomics

The next question moves from systems to society: what breaks when algorithms shape human outcomes? The growing reach of artificial intelligence, fintech platforms, and global digital systems introduces ethical pressures that industrial societies never faced at comparable scale. Automated decision systems (ADS) now influence hiring, lending, insurance pricing, content visibility, public safety assessments, and political engagement. These systems often generate unintended harms described in policy literature as *algorithmic externalities*: social, economic, or political costs imposed on individuals without their consent or participation. These harms are rarely deliberate. They arise from scale.

A hiring algorithm trained on historical resumes can reproduce gender or racial bias embedded in past labor markets. A college admissions model that weights legacy status or standardized test patterns can disadvantage first-generation applicants. A fintech platform offering instant credit can unintentionally trap vulnerable users in cycles of micro-debt. A ranking system that rewards engagement can amplify misinformation because outrage travels faster than nuance.

A well-documented example emerged in 2019, when researchers from the University of California, Berkeley, the University of Chicago, and Brigham and Women's Hospital analyzed a health risk prediction algorithm used across major American healthcare systems. The model evaluated more than 200 million patients and estimated future medical need using historical healthcare spending as a proxy. Because Black patients had historically received less medical care due to unequal access, their recorded spending was lower. The algorithm therefore inferred that they were healthier, even when clinical data showed higher rates of chronic illness. As a result, Black patients were referred to high-risk care programs at roughly half the rate of equally sick white patients. The system reproduced inequality automatically, despite no malicious intent by its designers.

As these systems scale, accountability becomes fragmented. Developers point to training data. Companies point to model outputs. Users encounter decisions shaped by processes they cannot inspect, challenge, or opt out of. Traditional consent frameworks fail because participation in modern life increasingly requires interaction with predictive systems. Avoiding algorithmic evaluation often means exclusion from essential services.

Ethical safeguards such as oversight boards, bias audits, and transparency requirements can mitigate these effects only when they are embedded early. Retrofitting ethics after deployment rarely corrects structural distortions that have already propagated through millions of decisions. Without deliberate governance, *Techonomics* does not merely reflect existing inequalities. It accelerates and amplifies them through automated scale.

The Role of International Standards

The final question of this chapter is pragmatic rather than philosophical: what prevents systemic failure in a world where technology crosses borders but law does not? The most effective stabilizing force is not national regulation alone, but technical harmonization. International standards provide shared rules of operation for systems that must function across jurisdictions even when political authority remains fragmented.

Organizations such as the International Organization for Standardization (ISO), the Basel Committee on Banking Supervision, and the Institute of Electrical and Electronics Engineers (IEEE) develop frameworks that shape cybersecurity practices, financial risk controls, identity formats, and artificial intelligence documentation. These bodies do not possess enforcement power. Their influence comes from necessity. Global systems cannot operate without shared technical assumptions, and institutions adopt these standards because fragmentation would make coordination impossible.

Harmonization reduces the friction created by conflicting national laws. A crypto exchange based in Singapore but serving customers in Europe and the United States may face different regulatory obligations in each region. Technically, however, it can rely on standardized methods for cryptographic key management, anti money laundering reporting formats, transaction logging, and security controls. Traditional finance has long depended on this model. The global banking system operates through the Society for Worldwide Interbank Financial Telecommunication (SWIFT), a standardized messaging network that allows thousands of financial institutions to communicate reliably across borders. Without it, cross-border payments would slow or fail entirely.

Artificial intelligence governance is beginning to follow a similar path. The European Union's AI Act, guidance frameworks in the United States, and China's Personal Information Protection Law (PIPL) reflect different political philosophies, yet practical alignment appears through shared concepts. Risk classification, documentation requirements, human oversight expectations, and auditability recur across jurisdictions. These overlapping principles function as a diplomatic layer for the digital era, allowing cooperation even when national policies diverge.

The alternative is fragmentation. Without harmonized standards, identity systems fail at borders, stablecoins behave unpredictably across markets, and cloud infrastructure becomes incompatible between regions. Innovation may still occur, but it cannot scale safely. Technical divergence becomes structural instability.

International standards cannot eliminate risk, nor can they resolve every governance dispute. What they provide is a common language for trust in a system where code, data, and value move freely while law remains territorial. In a borderless technological economy, shared standards are not optional. They are the minimum condition for stability.

Chapter X

The Knowledge Gap

People Unaware of AI and Crypto Opportunities

In the modern digital economy, knowledge functions as a form of capital as tangible as money or property. Awareness determines access. Individuals unfamiliar with artificial intelligence, blockchain, or cryptocurrency are not merely missing optional tools. They are increasingly excluded from economic participation itself. Digital literacy has shifted from a helpful advantage to a basic requirement for navigating finance, employment, entrepreneurship, and even digital identity. Familiarity with decentralized finance, algorithmic investing, and AI-driven productivity tools now plays a direct role in determining who advances and who falls behind. What emerges is a new form of stratification often described as *cognitive inequality*, a divide rooted not in inherited wealth but in awareness and comprehension.

This gap spans the entire socioeconomic spectrum. At the upper end, corporations accumulate Bitcoin (BTC), integrate generative AI into internal workflows, and deploy machine learning systems that accelerate productivity and compound competitive advantage. Capital grows more quickly in the hands of those who understand the technology early. At the opposite end, adoption occurs in informal or unregulated spaces. Individuals excluded from traditional finance use crypto for rapid payments, pseudonymous income, online community tokens, or speculative digital currencies tied to internet culture. The same underlying technologies empower billion-dollar enterprises and small decentralized networks, yet the outcomes differ sharply depending on intent, understanding, and strategic use.

Generational behavior widens this divide further. Younger users raised inside digital environments adopt new platforms with little hesitation. They experiment with AI assistants, automated investing tools, and decentralized marketplaces intuitively. Older generations, even when financially secure, often approach these systems with skepticism or confusion. Delay carries a cost. Early adopters accumulate disproportionate benefits, while late entrants encounter saturated markets, reduced upside, and higher barriers to participation.

Fear and misinformation intensify this hesitation. Headlines focused on crypto volatility or AI-driven job displacement discourage engagement altogether. Many assume these systems are too technical, too speculative, or too dangerous to approach responsibly. Meanwhile, individuals who explore these tools early gain access to new income streams, alternative career paths, and novel forms of digital ownership. The opportunity cost of inaction becomes visible only in hindsight. Those who dismissed early Bitcoin, overlooked Ethereum's rise, or ignored the commercial trajectory of AI often recognize too late how rapidly advantage compounded elsewhere.

The consequences extend beyond personal wealth. At a societal level, slow adoption of transformative technologies deepens inequality. Financial inclusion increasingly depends on familiarity with digital platforms, cryptographic systems, and algorithmic decision-making. Those who remain unaware risk exclusion from an economy where payments, credit, identity verification, and employment are mediated by AI-driven systems and decentralized networks. As awareness becomes a prerequisite for participation, the divide between insiders and outsiders grows more pronounced.

Awareness is not passive. It is economic agency. Understanding artificial intelligence, crypto assets, and modern fintech allows individuals to engage with expanding markets rather than remain subject to them. Ignorance narrows earning potential, weakens negotiating leverage, and limits the ability to navigate an economy that increasingly rewards digital competence. In the *Techonomic* era, knowledge is not simply information. It is the currency of mobility, shaping who gains access to opportunity and who remains structurally constrained.

Traditional Businesses vs. Tech Startups

Traditional businesses, shaped by decades of physical infrastructure and procedural design, increasingly struggle in an economy defined by speed, automation, and data-driven decision-making. Their systems were built for predictable markets, long planning horizons, and hierarchical control. These structures once provided stability. In a landscape where competitive advantage depends on rapid iteration, continuous deployment, and algorithmic optimization, they now impose constraints. Management scholars describe this burden as *legacy drag*, the operational friction that arises when institutions designed for industrial conditions attempt to function within digital paradigms.

Technology startups operate under a different logic. They are constructed for acceleration rather than preservation. Cloud-native infrastructure, modular software, and automated workflows allow them to test ideas quickly, scale rapidly, and update products continuously. Companies such as Stripe, Square, and Plaid did not simply introduce new financial products. They redesigned the architecture of financial services itself. By embedding automation and data analysis into onboarding, pricing, fraud detection, and customer support, they achieved efficiencies that traditional firms struggle to replicate through human-mediated processes and layered approval chains.

The contrast now appears across nearly every major industry. Legacy automakers spent decades refining physical production lines, yet companies such as Tesla approached the automobile as a software platform. Over-the-air updates, autonomous driving systems, AI-optimized energy management, and continuous data feedback reshaped competition in a sector long dominated by scale manufacturing. In banking, institutions constrained by compliance-heavy branch networks and legacy core systems now compete with neobanks that open accounts in minutes, settle payments instantly, and analyze financial behavior algorithmically. These disruptions occur not because incumbents lack capital or expertise, but because the structure of innovation itself has changed.

Organizational culture deepens this divide. Bureaucratic approval processes, risk aversion, and rigid hierarchies discourage experimentation within established firms. Incentive structures often reward short-term stability over long-term transformation, making meaningful change difficult to justify internally. Startups invert this logic. Iteration is expected. Failure is tolerated when it produces insight. Systems are built with the assumption that constant adjustment is normal rather than exceptional. Over time, this cultural difference becomes an economic one, determining which organizations adapt to emerging markets and which are forced into reactive positions.

At a deeper level, the shift is conceptual. The industrial economy rewarded firms that accumulated physical capital, controlled supply chains, and expanded distribution networks. The tech economy rewards firms that accumulate digital capital, harness network effects, and scale through software rather than factories. The companies reshaping global markets are not necessarily those with the largest workforces or physical footprints. They are those with the richest data, the most refined algorithms, and the most efficient computational environments.

Traditional firms are not destined to fail. Some successfully modernize by shedding legacy systems, rearchitecting operations, and integrating AI-driven decision-making. Those that do can remain competitive. Those that cannot adapt risk gradual irrelevance. In markets governed by speed, learning, and automation, adaptability is no longer optional. It is the dividing line between institutions that endure and those that quietly recede.

Why Old Models Are Sinking Ships

Traditional businesses built on industrial-era logic increasingly find themselves unprepared for the demands of a digital-first economy. Their organizational structures were designed for routine processes, predictable supply chains, and information that moved slowly through managerial hierarchies. These characteristics once produced resilience. In the current environment, they produce fragility. Economists describe this condition as *organizational inertia*, the tendency of established institutions to resist change even when adaptation is necessary for survival.

These firms were optimized for a world of paperwork, physical distribution, and annual planning cycles. The modern market operates on real-time feedback, continuous optimization, and systems that learn from data as events unfold. Startups face none of these constraints. They are built around cloud-native infrastructure, automated workflows, modular codebases, and AI-driven decision engines that adjust faster than traditional committees can convene. Their advantage is not novelty alone. It is structural.

This difference becomes visible through repeated case studies. Blockbuster recognized the potential of digital streaming early yet failed to reorganize around a model that threatened its existing revenue streams. Taxi monopolies across the world were displaced by ride-sharing platforms that combined GPS data, mobile payments, and algorithmic routing to deliver a superior experience at scale. Large retailers struggled against Amazon not because they lacked stores or logistics networks, but because Amazon constructed a computational system that integrated inventory, pricing, consumer behavior, and fulfillment into a single adaptive engine.

Organizational culture reinforces these outcomes. Traditional firms often rely on extended approval chains, fear-driven management, and incentive systems that punish deviation from established practice. In such environments, proposing change can appear riskier than maintaining outdated methods. Startups reverse this dynamic. Experimentation is expected, failure is treated as information, and speed becomes a strategic asset rather than a threat to stability. Institutions that cannot adopt this mindset find themselves perpetually responding to disruption rather than shaping it.

Beneath these examples lies a deeper economic transition. In the industrial era, value was tied to physical capital: factories, fleets, and distribution infrastructure. In the *Techonomic* era, value is tied to digital capital: datasets, algorithms, cloud capacity, and platform effects. Firms scale not by expanding physical presence, but by extending computational reach. A traditional company may spend years opening new locations. A digital company can serve millions of users instantly by replicating software.

The result is a widening innovation gap. Organizations that modernize their architecture, integrate AI into decision-making, and build around continuous learning gain compounding advantages. Those that delay become structurally disadvantaged, even when they begin with more capital, more employees, or stronger brands. The old models are not failing because they are mismanaged. They are failing because the economic environment they were designed for no longer exists.

Education and Awareness Divide

In the modern economy, access to knowledge has become one of the strongest predictors of economic mobility. Understanding artificial intelligence, blockchain systems, and modern fintech is no longer a niche specialization. It is an entry requirement for full participation in the emerging digital landscape. Individuals who can navigate algorithmic systems, manage digital assets, and interpret new financial tools gain measurable advantages over those who cannot. Education now functions as a form of economic power, shaping who benefits from technological progress and who is quietly pushed aside.

Traditional schools and universities struggle to keep pace with the speed of change. Curricula often lag years behind industry reality, leaving students underexposed to concepts such as decentralized finance, machine learning, digital identity systems, and cloud-native business models. Foundational subjects remain necessary, but practical familiarity with the systems shaping the next decade is frequently absent. The result is a generation that may be credentialed but strategically unprepared, reinforcing a cycle in which opportunity concentrates among those already embedded in digital environments.

The Knowledge Gap

A major driver of this inequality is what can be described as the *digital literacy gradient*, the uneven distribution of technical fluency across regions, households, and institutions. Students with reliable devices, high-speed internet, and early exposure to emerging technologies develop intuitive familiarity with AI tools, crypto interfaces, programming environments, and data platforms. Those without this access remain confined to instructional models designed for an economy that has already moved on. Technology becomes both ladder and barrier, accelerating advancement for some while structurally limiting others.

Digital platforms intensify this divide. Search engines, recommendation systems, and AI learning tools reward users who already know how to ask precise questions, evaluate sources, and synthesize information. A student who understands how to prompt AI systems, explore blockchain documentation, or troubleshoot code acquires knowledge at an accelerating rate. Another student, relying solely on traditional instruction, advances more slowly despite equal intelligence or effort. Digital learning environments amplify existing advantages rather than neutralizing them.

Alongside formal education, a parallel learning ecosystem has emerged. Self-directed learners turn to coding bootcamps, open-source AI models, decentralized finance dashboards, online academies, and tutorial platforms. These resources lower barriers to entry and allow motivated individuals to bypass traditional gatekeepers. Yet even this democratization introduces new stratification. Time, curiosity, stable access to technology, and freedom from immediate financial pressure determine who can fully exploit these tools. Opportunity remains unevenly distributed.

The effects are visible across labor markets. Economies are separating into digital insiders and digital outsiders. Those fluent in emerging systems experience faster wage growth, broader career mobility, and access to high-return opportunities. Those without this fluency encounter stagnant earnings, limited exposure to growth sectors, and increasing vulnerability to automation. Education becomes both protection against displacement and a lever for upward momentum.

Ultimately, education determines participation in *Techonomics.* Knowledge is no longer static. It is cumulative, iterative, and directly tied to economic value. Curiosity, continuous learning, and early exposure shape long-term outcomes. In an economy where information itself carries financial weight, the divide between the educated and the unaware becomes a divide in opportunity, wealth, and influence.

Winners and Losers in the Knowledge Economy

In the *Techonomic* era, the line between winners and losers is increasingly drawn by knowledge and adaptability rather than by tenure, physical capital, or geography. Early adopters of artificial intelligence, blockchain systems, and modern fintech capture disproportionate opportunities simply by recognizing value before it becomes obvious. Awareness itself becomes leverage. Those who understand emerging tools early accumulate advantages that compound over time, while late entrants face saturated markets and higher barriers.

This pattern is visible among technology entrepreneurs, AI engineers, and crypto innovators. Small teams with deep technical understanding can build platforms, protocols, or models that reshape entire industries. They do not merely compete within existing markets. They create new economic architectures that legacy firms struggle to follow. In contrast, workers in roles defined by routine or repetitive tasks face mounting pressure from automation. Algorithms, robotic systems, and machine learning models increasingly replace functions once performed by human labor, while retraining systems and labor protections lag behind the pace of change.

A widening asymmetry has emerged between those who can interpret digital systems and those who remain passive users of technology. Someone who understands how blockchain incentives function, how AI models are trained, or how data pipelines operate can perceive risks and opportunities invisible to users who interact only with polished consumer interfaces. This produces a form of epistemic inequality, where understanding itself acts as capital. Insight opens doors long before financial returns become visible to the broader public.

Even within the same profession, knowledge determines position and earning power. In healthcare, practitioners who learn to work alongside diagnostic AI, predictive analytics, and automated workflows advance faster than those who resist integration. In finance, professionals who master algorithmic trading, risk modeling, or digital asset structures outpace peers anchored to older methods. The gap does not arise from intelligence or effort alone. It arises from proximity to evolving systems.

Across the broader economy, knowledge increasingly behaves like capital. Skills such as data analysis, computational thinking, financial literacy, and technical fluency translate directly into economic influence. Digital competence is no longer a differentiator. It is a baseline requirement. Those who invest in continuous learning and experimentation gain access to expanding markets. Those who do not find their options narrowing, even when they possess experience or credentials earned under previous economic conditions.

The societal implications are significant. As returns on knowledge compound, inequality widens. Public debate increasingly centers on education reform, digital access, and reskilling initiatives as governments attempt to slow divergence between digital insiders and outsiders. Yet the knowledge economy also creates new paths for mobility. Individuals with limited traditional credentials can bypass established gatekeepers, build online businesses, or enter high-growth sectors by mastering emerging tools.

The *Techonomic* era ultimately rewards awareness, adaptability, and skill acquisition. Those who learn early shape markets, capture opportunity, and influence outcomes. Those who delay or disengage become increasingly marginalized, not through exclusion, but through structural disadvantage. In a world where information itself carries economic weight, knowledge becomes the dividing line between relevance and obsolescence.

Chapter XI

Social Upheaval

Job Displacement and Automation Risks

Automation is no longer merely replacing discrete tasks. It is reshaping the social fabric that once anchored economies and communities. Entire categories of work that provided identity, stability, and regional pride are shrinking, leaving millions uncertain not only about income but about purpose in a world where human contribution feels increasingly negotiable. This emotional fallout is now widely described as *technological displacement anxiety*, the growing sense that one's skills, status, and future are being eroded by forces far beyond personal control.

The deepest disruption appears in communities tied to occupations now threatened by intelligent systems. Towns built around transportation, manufacturing, customer service, or administrative work experience a form of slow erosion. As employers automate logistics networks, modernize warehouses, or deploy AI-based clerical tools, surrounding communities hollow out. Small businesses close, local tax bases shrink, and younger residents migrate toward technology-centered regions. What begins as an economic shift becomes cultural and psychological decline, leaving behind populations that feel increasingly disconnected from national prosperity.

This transformation also reshapes personal identity. For many workers, employment was not merely a source of income but a measure of dignity, contribution, and social belonging. When machines begin to outperform people in roles once considered skilled or respected, individuals confront a sense of redundancy that retraining alone cannot resolve. Status loss becomes deeply personal. Resentment often follows, directed toward institutions, elites, or technologies perceived as indifferent to lived experience. This identity disruption frequently feeds political polarization, as people search for leaders or narratives that validate their frustration.

Generational divides intensify the strain. Younger workers often integrate AI tools seamlessly into their workflows, gaining productivity and wage advantages. Older workers, even when experienced and capable, may struggle to adapt to systems that evolve faster than traditional training models allow. This divergence creates tension not only in workplaces but within families, accelerating the sense of social fragmentation.

Automation also reshapes the geography of opportunity. Prosperity concentrates around technology hubs where skilled labor, capital, and infrastructure cluster. Regions dependent on automatable work enter prolonged decline, fostering competing narratives of progress. One celebrates technological efficiency and innovation. The other mourns the erosion of stability, tradition, and communal identity. Distrust grows between these worlds, fueling movements that reject not only specific technologies but the broader direction of economic change.

The challenge, then, is not simply that machines replace work. It is that rapid automation destabilizes the foundations of social life: identity, purpose, community, and the expectation that effort reliably produces opportunity. Societies that fail to address this rupture risk entering an era where technology advances steadily while the culture surrounding it fractures, leaving large segments of the population feeling abandoned in a system they no longer recognize.

Government Intervention vs. Laissez-Faire Collapse

As technological disruption accelerates, governments face a defining choice: intervene to manage mounting social and economic instability or allow markets to evolve with minimal oversight and accept the risk of widespread dislocation. Policy tools range from targeted taxation and labor protections to AI auditing regimes, digital identity frameworks, and large-scale retraining programs. In principle, governments possess the authority to shape how automation, artificial intelligence, and fintech influence society. In practice, many lack the institutional capacity to act at the speed these technologies demand.

This constraint is increasingly described as a *regulatory capacity gap*, the widening distance between technological change and the ability of public institutions to understand, oversee, and respond to it. When innovation unfolds in months while legislation evolves over years, governance becomes reactive rather than protective. Societies absorb disruption first and attempt regulation later, often after damage has already spread across labor markets, communities, and public trust.

Under-regulation exposes clear risks. Without guardrails, wealth and influence concentrate around early adopters, dominant platforms, and capital owners, while the broader workforce confronts stagnant wages, declining job security, or outright displacement. Personal data becomes an extractable commodity with minimal constraint. Digital labor is mediated through gig platforms that shift risk onto workers. Algorithmic systems shape access to credit, employment, and opportunity with limited transparency or recourse. As these pressures accumulate, public confidence erodes and resentment grows. Left unchecked, laissez-faire conditions can provoke unrest similar to earlier periods of industrial upheaval, when citizens demanded protection from forces they could neither negotiate with nor escape.

Excessive intervention introduces a different set of dangers.

Heavy-handed restrictions on AI development, fintech innovation, or decentralized networks can suppress entrepreneurship and delay the adoption of genuinely beneficial technologies. Rigid compliance regimes, inflexible identity requirements, or slow-moving supervisory bodies may suffocate early-stage innovation before it matures. Capital and talent migrate toward more permissive jurisdictions, leaving stricter regions with diminished influence over systems that continue to evolve elsewhere. Regulation that ignores technological realities risks becoming self-defeating.

Recent governance experiments reflect an attempt to navigate this balance. AI oversight boards, digital taxation proposals, and specialized fintech supervisory agencies seek to monitor systemic risk while preserving innovation. These efforts acknowledge that traditional regulatory models, designed for slower industrial cycles, are insufficient in a world governed by algorithmic scale and machine-speed markets. To remain effective, governance must evolve from static rulemaking toward adaptive frameworks capable of shaping innovation rather than merely responding to its consequences.

One emerging approach is *anticipatory governance*, which uses predictive modeling to identify risk before it crystallizes into crisis. Instead of waiting for mass unemployment, regional decline, or financial instability, governments analyze automation trends, labor displacement patterns, and market signals to intervene earlier. Predictive analytics can indicate where disruption is likely to concentrate and which communities require support. While still experimental, this approach reflects a recognition that reactive governance is no longer viable.

Global coordination has become equally important. Digital firms operate across borders, making purely national regulation easy to evade through relocation. In response, countries are experimenting with shared frameworks such as the Organisation for Economic Co-operation and Development (OECD) digital taxation agreements and cross-border AI evaluation standards designed to reduce regulatory arbitrage. As technology globalizes, governance must follow, or risk losing relevance while instability persists.

The state's role in the *Techonomic* era is therefore decisive. Failure to intervene wisely risks economic fragmentation, political instability, and loss of public legitimacy. Overcorrection risks stifling the innovation that generates societal wealth. The challenge is not merely technical. It is ethical and strategic. Stability in an automated world requires institutions capable of adapting as quickly as the systems they seek to guide.

Universal Basic Income

Universal Basic Income (UBI) has emerged as one of the most debated responses to economic disruption driven by automation and artificial intelligence. Its premise is straightforward: provide every citizen with a regular, unconditional payment sufficient to guarantee a minimum standard of living, regardless of employment status. Advocates argue that UBI could stabilize household income, reduce poverty, and give individuals the freedom to pursue education, retraining, entrepreneurship, or creative work without constant financial pressure. In a labor market increasingly shaped by technological displacement, UBI is often presented as a buffer against volatility rather than a replacement for work.

Early pilot programs offer limited but instructive evidence. Finland's nationwide experiment with unemployed participants reported improvements in mental well-being, reduced stress, and smoother reintegration into the labor force, even though long-term employment effects were modest. Similar trials in Canadian provinces and several U.S. cities showed increased consumption stability and lower financial anxiety, while also revealing the immense fiscal burden required to sustain such programs at scale. These experiments suggest that UBI can soften immediate hardship but leave unanswered questions about its long-term macroeconomic effects.

Implementation remains the central challenge. Funding a universal stipend requires higher taxation, significant public borrowing, or the restructuring of existing welfare systems. Critics warn that injecting large sums of unconditional income could fuel inflation and erode the purchasing power UBI is intended to protect. Others raise cultural concerns, arguing that unconditional payments may weaken social norms around work, contribution, and responsibility. More fundamentally, UBI does not directly address structural issues such as corporate concentration, unequal access to education, or the distribution of ownership in automated systems.

Interest in UBI has intensified because the foundations of wealth creation are changing. For decades, households accumulated wealth indirectly through stable employment and broad participation in pension funds, index funds, and retirement systems. That model assumed relatively predictable labor demand and incremental productivity growth. In the 2020s, this assumption weakened. Digital assets began exhibiting extreme volatility. AI-driven trading systems identified patterns beyond human perception and executed strategies at machine speed. Returns increasingly flowed toward institutions with advanced predictive infrastructure, leaving ordinary participants structurally disadvantaged.

At the same time, job security eroded. Routine digital labor once considered stable, including administrative work, customer support, and content production, became vulnerable to automation. Value creation shifted away from human judgment toward automated decision engines. When employment stability and investment opportunity weaken simultaneously, the appeal of a guaranteed economic baseline grows stronger.

These conditions raise a deeper question: who should administer a universal income in an era shaped by AI-driven markets and decentralized finance? Should responsibility remain with national governments, shift toward regional authorities, or evolve into new forms of digital public infrastructure capable of distributing funds transparently and efficiently? As financial systems become increasingly automated, the question of administration becomes inseparable from the question of legitimacy.

Alternative approaches attempt to preserve stronger incentives to work. Negative income tax models provide support only below certain income thresholds. Conditional programs link assistance to employment, training, or educational participation. These systems maintain clearer ties between contribution and support but introduce administrative complexity and may struggle to respond effectively to widespread technological unemployment. The philosophical tension remains unresolved. Should economic security be unconditional, or should it remain contingent on measurable participation?

UBI is best understood as one instrument within a broader set of responses to automation-driven disruption. On its own, it cannot compensate for weak education systems, limited access to retraining, or deep disparities in digital literacy. Without complementary reforms, UBI risks becoming either fiscally unsustainable or socially insufficient. As the *Techonomic* era accelerates, societies must consider both immediate safety nets and long-term strategies that ensure technological progress expands opportunity rather than concentrating it further.

Corporate Control Over Society

In the *Techonomic* era, a new concentration of power has emerged. Large technology and financial corporations now exert influence over communication, commerce, information flow, and financial access at a scale once associated primarily with governments. A small number of platforms shape what billions of people see, how transactions are processed, and which voices are amplified or suppressed. Economic power increasingly overlaps with social control, producing what scholars describe as *platform sovereignty*, the ability of private corporations to establish rules, enforce norms, and govern behavior without formal democratic authority.

Corporate governance now fills spaces traditionally occupied by public institutions. Major platforms determine access to financial tools, set content standards, define acceptable speech, collect personal data, and deploy artificial intelligence systems that influence opportunity and visibility. These decisions are made by corporate leadership rather than elected bodies and often operate beyond meaningful public oversight. As a result, governance shifts from law to policy, and from public deliberation to internal corporate decision-making. The distinction between regulation and product design becomes increasingly blurred.

This concentration of influence extends beyond consumer

technology. Large asset managers such as BlackRock, Vanguard, and State Street exert structural control across global markets. BlackRock alone manages over twelve trillion dollars in assets, exceeding the gross domestic product of most nations. Through index funds and institutional holdings, it maintains ownership stakes across a vast portion of the global economy. Its internal risk management platform, Aladdin, provides analytics used by banks, pension funds, and governments worldwide. In effect, BlackRock does not merely participate in markets. It helps define how risk is measured, managed, and priced across entire financial systems.

Such financial concentration translates into governance power. Through shareholder voting and advisory influence, a small number of asset managers shape corporate decisions related to executive compensation, environmental policy, political engagement, and long-term strategy. These interventions are often framed as fiduciary responsibility, yet they centralize decision-making authority in entities that operate with limited transparency and minimal public accountability. Influence once exercised through legislation increasingly flows through ownership structures and algorithmic risk models.

Technology platforms mirror this dynamic in the cultural and informational sphere. Companies such as Apple, Amazon, Meta, Google, and emerging AI and crypto firms operate across jurisdictions, frequently beyond the effective reach of national regulators. Their scale allows them to establish global norms for commerce, identity, communication, and content moderation. They influence political discourse, consumer behavior, and financial inclusion not through coercion, but through architecture. The design of interfaces, ranking systems, and recommendation algorithms quietly determines what is visible, viable, or profitable.

The implications reach into the foundations of public life. Democratic participation, free expression, and economic mobility increasingly depend on decisions made within private systems rather than public institutions. Algorithms determine which ideas circulate, which creators are rewarded, and which opportunities appear before users. Digital markets no longer resemble open competitive fields. They function as curated environments shaped by proprietary data, ranking logic, and behavioral optimization. While these systems deliver efficiency and convenience, they also narrow individual autonomy and raise fundamental questions about consent and control.

Power is therefore shifting from elected governments toward private entities at a pace without historical precedent. The central issue is no longer whether technology generates wealth, but whether societies can preserve autonomy when access, opportunity, and information are governed by corporations operating beyond democratic constraint. As corporate power expands, citizens and institutions must confront the reality that authority in the *Techonomic* era increasingly resides outside traditional public structures. The challenge ahead is not only economic. It is political and civic, requiring a rethinking of governance in a world where private infrastructure mediates public life.

Social Stability in a Tech-First World

The rapid advance of artificial intelligence, automation, and digital platforms is reshaping societies faster than their institutions can adapt. Rising inequality, job displacement, and the concentration of corporate power place sustained pressure on social cohesion. Economic opportunities increasingly depend on digital fluency, and as a result, the gap between those who can navigate technological systems and those who cannot becomes a source of tension. Stability in this environment is no longer guaranteed by growth alone. It requires deliberate social adaptation.

Societies now face a narrowing window for peaceful adjustment. History offers warnings. The Industrial Revolution produced riots and unrest when mechanization advanced faster than labor protections. Late twentieth-century globalization fractured communities left behind by deindustrialization. The *Techonomic* era intensifies these pressures because innovation spreads globally and instantaneously, leaving little time for gradual absorption. When disruption arrives everywhere at once, its social consequences compound.

Education and retraining are central to maintaining cohesion, yet they must evolve beyond traditional models. Workers displaced by automation require pathways into new forms of economic participation, not symbolic credentials disconnected from market demand. Education systems must emphasize adaptability, digital literacy, and continuous learning rather than static expertise. Without these reforms, retraining efforts risk becoming procedural responses to structural change rather than genuine solutions.

Public trust is a fragile variable in this transition. When citizens perceive that technological systems primarily benefit corporations and elites, confidence in institutions erodes. Algorithmic decisions that affect employment, credit, or access to services often appear opaque and unchallengeable. When accountability is unclear, individuals experience technology as imposition rather than opportunity. This perception accelerates disengagement and fuels political polarization.

Psychological strain magnifies these effects. Sociologists describe a condition of *technological disequilibrium*, where the pace of innovation exceeds a community's ability to integrate change culturally or emotionally. People struggle not only with economic uncertainty, but with the feeling that progress is occurring without them. In such environments, nostalgia, reactionary politics, and resistance movements gain traction, offering emotional certainty where technological systems provide none.

Global examples underscore these dynamics. Automation has

triggered protests in manufacturing regions. Algorithmic management has prompted labor strikes in logistics and delivery sectors. Generational divides widen as younger, digitally fluent workers advance more rapidly than older cohorts. These patterns reveal that social instability is not an accidental byproduct of technological change. It is a predictable outcome when adaptation lags innovation.

Long-term stability in a tech-first world depends on renewing the social contract. Societies must align technological progress with equitable access, meaningful work, and civic participation. Without such alignment, innovation risks deepening fragmentation rather than expanding opportunity. With it, the disruptive forces of *Techonomics* can be directed toward resilience rather than rupture. The challenge is not resisting change but shaping it before instability becomes irreversible.

Chapter XII

Learning Rewired

The Limits of Traditional Education

Traditional education systems were not designed organically. They were engineered. The modern school structure traces its origins to the nineteenth-century Prussian model of education, a system developed to produce disciplined, literate citizens capable of serving an industrial state. Its goals were uniformity, obedience, and reliability. The bell schedule, age-based cohorts, standardized curricula, grading systems, and teacher-centered classrooms were mechanisms of coordination, not curiosity. They trained students to follow instructions, respect hierarchy, and perform repeatable tasks within bureaucratic institutions.

This model proved effective in an industrial economy that valued predictable labor and slow, incremental change. Factories, clerical offices, and state administrations required workers who could execute predefined roles with consistency. Education mirrored that need. Compliance was rewarded. Deviation was corrected. Mastery meant repetition.

That logic no longer aligns with the conditions of the *Techonomic* era. Artificial intelligence, blockchain systems, decentralized work, and rapid digital experimentation have replaced stability with constant change. Yet schools continue to reward memorization and procedural conformity at the very moment when information itself has become abundant and inexpensive. Colleges extend this contradiction. Lectures and examinations still emphasize information storage even as search engines and generative AI systems retrieve facts, summarize research, and identify patterns faster and more accurately than any human. The value of education has shifted away from recall and toward interpretation, judgment, and decision-making under uncertainty.

The growing mismatch between education and reality reflects a form of cognitive lag. Institutions optimized for the industrial century now operate inside a digital economy that evolves faster than curricula can be revised. Students are prepared for roles that no longer exist while being undertrained for the environments they will actually face.

Even professions once considered insulated from disruption reveal this gap. Lawyers now rely on automated contract analysis. Radiologists work alongside AI systems that detect anomalies with machine precision. Financial analysts depend on algorithmic engines that process markets beyond human cognitive speed. Yet many academic programs still teach these disciplines as if human labor will remain the primary engine of analysis. The future belongs less to those who can follow established procedures and more to those who can contextualize outputs, question automated conclusions, and apply judgment where machines reach their limits.

The Prussian model also treats students as passive recipients rather than active problem solvers. Standardized curricula discourage exploration and penalize deviation. Learning becomes linear, controlled, and externally motivated. This structure prepares students for institutions built on command and control, not for environments defined by experimentation, iteration, and ambiguity. In a tech economy, mental flexibility and adaptive reasoning are foundational. Graduates who lack these capacities enter the workforce structurally behind from the outset.

Education has further failed to adapt to a world where learning does not end with early adulthood. Careers now evolve continuously. Tools change, industries reorganize, and entire fields rise and fall within a single decade. The assumption that education is completed before professional life begins belongs to the industrial century. In the *Techonomic* era, continuous learning is the minimum requirement for relevance.

If machines can store, retrieve, and process information on demand, human specialization must shift elsewhere. Critical reasoning, creativity, ethical judgment, and contextual understanding define the new foundations of human work. The Prussian model was engineered for stability and repetition. The *Techonomic* era demands adaptive thinkers capable of interpreting complexity, shaping new systems, and collaborating with intelligent tools rather than competing against them.

Daily Life as the Classroom

As formal education struggles to adapt, practical experience has quietly regained its historical role as a primary engine of learning. In the *Techonomic* era, competence is increasingly developed not through abstraction alone, but through engagement with real systems, real constraints, and real consequences. Everyday activities such as cooking, budgeting, maintaining a household, managing a small business, or repairing equipment cultivate the same cognitive capacities required to navigate complex digital environments. These activities train individuals to observe patterns, diagnose problems, allocate resources, and adapt to changing conditions. They develop judgment rather than recall.

This form of learning reflects what cognitive scientists describe as *embodied cognition*, the principle that understanding deepens when thinking is integrated with action. A garden teaches systems thinking as weather, soil, timing, and effort interact. A home repair teaches iterative troubleshooting and planning under uncertainty. Managing personal finances teaches foresight, risk assessment, and delayed gratification. These skills cannot be memorized or simulated convincingly. They are acquired through repetition, error, and lived experience.

The absence of this kind of learning has consequences beyond the individual. A generation educated primarily for administrative systems and corporate specialization was trained to recall information and perform narrow functions, not to maintain, steward, or repair the physical environments it inhabits. The result is visible everywhere in decaying infrastructure, neglected public spaces, and communities surrounded by disorder yet uncertain how to restore them. Education optimized for abstraction produced efficiency for institutions, but it weakened practical responsibility for the world immediately at hand.

Experiential learning also builds resilience. When individuals fix what breaks, negotiate outcomes, or navigate interpersonal conflict, they develop tolerance for ambiguity and failure. These traits translate directly into modern professional environments shaped by AI, automation, and constant iteration. A person accustomed to solving tangible problems adapts more easily to unfamiliar tools, evolving workflows, and roles that lack clear precedent.

Historically, apprenticeship systems embodied this logic. For centuries, young workers learned trades by observing masters, beginning with simple tasks and gradually assuming full responsibility. That structure transmitted not only technique, but judgment, ethics, and professional identity. The same dynamic persists today. A senior engineer mentoring a junior developer, a carpenter teaching a teenager to frame a wall, or a small business owner involving family members in inventory and bookkeeping conveys practical intelligence that no curriculum can formalize. In a world saturated with information, mentorship remains one of the few irreplaceable educational resources.

Households and family enterprises also function as informal learning institutions. Children who participate in maintaining property, growing food, repairing tools, or contributing to a family business observe entire systems from input to outcome. They learn accountability, sequencing, cooperation, and cause-and-effect reasoning. These environments cultivate competence through participation rather than instruction, producing confidence grounded in capability rather than credential.

Formal institutions are beginning, slowly, to recognize these limits. Project-based learning, simulated workplaces, and apprenticeship partnerships attempt to reintroduce reality into education. When students design, build, manage, or negotiate something that exists beyond an exam, learning becomes durable. Knowledge attaches to consequence.

In the *Techonomic* era, this shift is not nostalgic. It is structural. As intelligent systems absorb repetitive and informational labor, human value concentrates in initiative, judgment, creativity, and the ability to act under uncertainty. Daily life itself becomes an ongoing classroom, one that teaches lessons that no database can store and no algorithm can reproduce. Learning by doing is not an alternative to education. It is the foundation upon which adaptive intelligence is built.

AI and Personalized Learning

Artificial intelligence is altering education not by replacing teachers, but by dismantling the uniform model that defined industrial schooling. Traditional classrooms assume that students should move at the same pace, absorb the same material, and demonstrate understanding through identical assessments. AI systems reject that assumption. They treat learning as an individualized process shaped by pace, curiosity, and aptitude, allowing instruction to adapt continuously to the learner rather than forcing the learner to adapt to the system.

Adaptive learning platforms analyze how a student engages with material, where they hesitate, and which explanations lead to comprehension. Instead of advancing by age or calendar, learners advance by mastery. A student who struggles with a concept receives alternative explanations, additional examples, or slower progression. A student who excels moves forward without artificial delay. This replaces the batch-processing logic of classrooms with a feedback-driven model closer to how people actually learn.

AI also shifts the role of educators. When machines handle routine grading, content delivery, and basic assessment, teachers are freed from clerical labor and repositioned as mentors. Their value lies less in recitation and more in judgment, encouragement, interpretation, and ethical guidance. In a world where information is instantly accessible, the human task is not to transmit facts, but to help students learn how to evaluate sources, frame questions, and apply knowledge responsibly.

Personalized learning environments also enable what can be described as directed learning paths. A student interested in design can integrate visual arts, software tools, and storytelling. Another focused on entrepreneurship can combine finance, negotiation, and market analysis. A third drawn to engineering can emphasize logic, physics, and systems modeling. AI systems detect interests and performance patterns, adjusting material to reinforce motivation rather than suppress it. Learning becomes exploratory rather than compulsory.

Immediate feedback strengthens this process. When confusion appears, systems respond instantly rather than weeks later through exams. When competence emerges, difficulty increases accordingly. This dynamic pacing reduces boredom, frustration, and disengagement, problems that traditional education treats as behavioral issues rather than structural failures. Learning becomes iterative and responsive, mirroring the environments students will later encounter in professional life.

AI also expands access to advanced instruction once limited to elite institutions. Language learning systems simulate conversation partners. Coding assistants guide learners through real projects. Virtual laboratories allow experimentation without physical facilities. A student in a rural or underfunded region can now engage with material previously reserved for top-tier universities or corporate training programs. Geography loses its power as a gatekeeper.

Perhaps most significantly, AI enables education to continue beyond formal schooling. As careers shift and industries evolve, individuals no longer rely on periodic retraining or additional degrees. Instead, AI-driven systems function as ongoing learning companions, offering targeted instruction as new tools, regulations, or opportunities emerge. Education becomes continuous, adaptive, and integrated into working life rather than confined to youth.

The deeper implication is not technological but philosophical. When machines handle retrieval and repetition, human learning must emphasize interpretation, judgment, creativity, and ethical reasoning. AI does not eliminate the need for education. It clarifies its purpose. In the *Techonomic* era, personalized learning aligns education with reality, preparing individuals not to memorize static knowledge, but to navigate a world defined by constant change.

Specialization Without the Bottleneck

The traditional education model was built around scarcity. Knowledge was difficult to access, expertise was concentrated in institutions, and credentials served as proxies for capability. Multi-year degree programs made sense in a world where information moved slowly and professional roles remained stable for decades. In the *Techonomic* era, that structure has become a bottleneck. Technology evolves faster than universities revise curricula, and industries now demand skills that did not exist when students first enrolled.

In response, a more flexible model of specialization is emerging. Instead of requiring years of generalized coursework before practical engagement, learners increasingly pursue targeted, modular education designed around specific competencies. Short intensive programs, certifications, and focused training pathways allow individuals to acquire economically valuable skills in months rather than years. This accelerates entry into high-demand fields while keeping learning aligned with real market needs.

This approach places control back in the hands of the learner. Rather than following a rigid academic sequence, individuals assemble skill portfolios that reflect their goals and interests. Someone might begin with smart contract development, add financial analytics, and later study cloud security as industry requirements shift. These combinations produce what can be described as *microdomain mastery*, deep expertise in narrow but highly relevant areas that generate immediate value without requiring broad credential inflation.

Stackable credentials reinforce this shift. Instead of a single diploma summarizing years of study, learners accumulate verifiable proof of discrete capabilities. A candidate might demonstrate proficiency in prompt engineering, data pipeline design, and decentralized application security, each credential earned through focused instruction and applied work. Employers increasingly favor this evidence because it signals current, usable skill rather than abstract qualification.

This model also supports lifelong learning. As automation alters job requirements, workers can reskill without stepping away from employment or reentering full-time education. A logistics professional can study AI optimization tools. A marketer can learn data analytics. A finance worker can acquire knowledge of blockchain settlement systems. Learning becomes iterative and continuous, reducing the shock of displacement and increasing career resilience.

Educational institutions and private platforms are adapting to meet this demand. Bootcamps, online academies, and community training hubs emphasize applied projects, mentorship, and portfolio development. Learners demonstrate competence by building systems, analyzing data, or solving real problems rather than memorizing theory for exams. What matters is not how long someone studied, but what they can actually do.

One of the most important consequences of this shift is the erosion of traditional access barriers. A motivated learner no longer needs to relocate, incur massive debt, or pass through elite gatekeeping institutions to enter high-opportunity fields. With reliable internet access and disciplined study, individuals in small towns or underserved regions can acquire world-class skills and compete in global markets. Geography and background matter less when education is modular, remote, and skill-verified.

Specialization without the bottleneck does not eliminate the value of broad education. It reorders priorities. General knowledge provides context, but economic participation increasingly depends on adaptable expertise and continuous updating. In the *Techonomic* era, relevance decays quickly. Systems that allow individuals to specialize rapidly, pivot efficiently, and learn continuously are not merely educational innovations. They are survival mechanisms in an economy defined by speed, complexity, and constant change.

Redefining Work and Career Paths

In *Techonomics*, education and work are no longer sequential stages of life. They have merged into a continuous process in which learning,

earning, and adaptation occur simultaneously. Knowledge is acquired not only in classrooms but through direct engagement with evolving systems, tools, and markets. The boundary between preparation and participation has dissolved. Individuals now build competence while actively shaping their economic roles rather than completing education before entering productive life.

This shift alters what it means to be qualified. Mastery of a single discipline in isolation is no longer sufficient. Professionals must understand how technology interacts with regulation, finance, social norms, and institutional power. A technically skilled worker who lacks awareness of legal constraints, market incentives, or ethical consequences may build efficient systems that fail in practice. Conversely, individuals who grasp these intersections gain strategic judgment, allowing them to navigate complexity rather than merely execute tasks.

As automation advances, the nature of human work continues to migrate toward areas machines cannot fully replicate. Emotional intelligence, ethical reasoning, physical craftsmanship, leadership, negotiation, and contextual decision-making increase in value as algorithms assume routine analysis and optimization. These forms of work depend on judgment shaped by experience rather than calculation alone. They anchor human contribution in domains where meaning, trust, and responsibility matter as much as efficiency.

Central to this transformation is cognitive complementarity, the relationship between human judgment and machine capability. Artificial intelligence excels at pattern recognition, scale, and speed. It does not understand consequence, purpose, or moral weight. Humans interpret outputs, challenge assumptions, and decide when efficiency must yield to wisdom. Professionals who learn to question automated conclusions, contextualize data, and integrate human values into decision-making gain leverage across every sector, from healthcare and logistics to finance and creative production.

The workforce that emerges from this environment is hybrid by necessity. Digital literacy becomes as foundational as reading and writing. At the same time, philosophical reasoning, long-term thinking, and interpersonal skill remain indispensable. Workers must understand how AI systems function without surrendering authority to them. Technology becomes a collaborator rather than a master, enhancing human agency rather than replacing it.

Education systems increasingly reflect this convergence. Hybrid models combine AI-driven instruction, mentorship, apprenticeship, and experiential learning. Intelligent tools deliver personalized knowledge with speed and precision. Human mentors transmit judgment, discipline, and ethical orientation formed through experience rather than instruction alone. Applied work binds both to real consequences, where decisions carry cost, responsibility, and lasting effect. Together, these elements form an educational structure aligned with the realities of modern labor rather than the assumptions of an industrial past.

Careers themselves become fluid rather than fixed. Individuals move between roles, industries, and forms of work over time. Identity shifts from job title to capability. Stability comes not from permanence but from adaptability, the capacity to learn, unlearn, and relearn as conditions change. In such a world, professional resilience depends less on credentials and more on the ability to integrate new tools, interpret unfamiliar systems, and contribute meaningfully under uncertainty.

Work and education have therefore converged into a single adaptive cycle. Those who cultivate judgment, flexibility, and technological fluency will shape their own paths within it. Those who cling to rigid roles or resist new tools will find their options narrowing. The future belongs neither to machines nor to institutions alone, but to individuals capable of working with intelligent systems while retaining human authority over purpose, value, and direction.

Chapter XIII

Old vs. New Money

Technocrats vs. Industrial Oligarchs

The *Techonomic* era has produced a clear divide between two distinct forms of economic power. Industrial oligarchs built wealth through factories, land, commodities, and large physical infrastructure. Their fortunes were rooted in tangible assets that expanded slowly and predictably, often across multiple generations. Power accumulated through control of production, labor, and supply chains, and stability was considered a virtue. By contrast, *technocrats* derive wealth from data, algorithms, digital platforms, and network effects. Their assets are intangible, scalable, and capable of compounding at speeds unknown to industrial capitalism. This shift represents a structural inversion of wealth, where control over information routinely outperforms control over material production.

The cultural divide between these groups mirrors their economic foundations. Industrial elites traditionally valued discretion, continuity, and institutional longevity. Their influence flowed through banks, trade associations, political relationships, and inherited positions of authority. Technocrats operate according to a different logic. Speed, experimentation, and visibility are not liabilities but advantages. They disrupt entire sectors through software releases, platform redesigns, and algorithmic deployment rather than through ownership of land or machinery. Prestige in this environment is no longer tied to stewardship of physical capital, but to the ability to scale influence rapidly across digital systems.

History highlights the magnitude of this transition. Industrial titans such as Carnegie, Ford, Rockefeller, and Vanderbilt spent decades building empires based on steel, oil, railroads, and manufacturing. Their fortunes expanded in proportion to physical output and logistical reach. Growth was constrained by extraction rates, factory capacity, and transportation limits. In contrast, modern technocratic wealth grows through network leverage. Elon Musk influences aerospace, automotive manufacturing, and artificial intelligence simultaneously. Jeff Bezos constructed a commercial empire without expanding physical territory in the traditional sense. Mark Zuckerberg governs global communication flows that shape culture, politics, and social interaction. Their power scales through code and connectivity rather than through furnaces or assembly lines.

This transformation also alters the nature of power itself. Industrial oligarchs shaped societies indirectly by controlling employment, infrastructure, and resources. Technocrats shape behavior directly. They influence how people communicate, what information they encounter, how transactions occur, and how identity is verified. A minor change in a recommendation algorithm can alter political discourse. A modification to cloud services can affect national security priorities. A new AI model can redefine professional standards across entire industries. Control over information architecture now rivals control over physical capital as the primary source of influence.

The tension between old and new money appears across nearly every sector. Legacy banks rely on regulatory familiarity and institutional trust, yet increasingly lose ground to fintech firms that deploy automation, AI driven risk assessment, and decentralized settlement systems. Media conglomerates built on print and broadcast struggle against platforms that monetize attention in real time. Industrial firms optimize supply chains while technology firms reconstruct value chains altogether. The conflict is not simply competitive. It reflects incompatible assumptions about how value is created and sustained.

Public perception reinforces this divide. Old money is often associated with stability, restraint, and custodianship of tradition. New money appears volatile, ambitious, and culturally dominant. Technocrats accumulate public recognition alongside wealth, shaping narratives and public imagination, while industrial elites maintain influence quietly through established networks. These contrasting archetypes reveal a society negotiating between two competing models of authority and legitimacy.

At its core, the rivalry between technocrats and industrial oligarchs is not merely economic. It is a struggle over who defines the future. Industrial wealth reflects a world organized around physical capital and incremental growth. Technocratic wealth reflects a world organized around digital leverage and continuous reinvention. As *Techonomics* advances, influence increasingly belongs to those who can adapt to rapid technological change, while those anchored to twentieth century frameworks confront the limits of systems that once guaranteed their dominance.

Abandoned Factories and Digital Empires

The global economy has crossed a decisive threshold. Physical industry no longer anchors national prosperity in the way it once did. Across large portions of the developed world, steel mills, textile plants, and automobile factories that once sustained entire regions now stand silent. These structures are not merely relics of an earlier economy. They mark zones of displacement where industrial activity declined faster than communities could adapt, leaving behind economic and social vacuums that persist for generations.

In contrast, digital empires rise on assets that are largely invisible. Companies such as Amazon, Meta, Tesla, and a growing constellation of technology firms generate value from cloud infrastructure, software ecosystems, data pipelines, and algorithmic coordination. Their growth does not require dense physical footprints or proximity to raw materials. Instead of smokestacks and rail yards, they build server clusters, development platforms, and global user networks that scale across borders with minimal marginal cost. Economic power becomes detached from geography, altering how regions compete and how prosperity concentrates.

The human consequences of this realignment are severe. Communities once organized around manufacturing lose more than jobs. They lose identity, continuity, and social cohesion. Local businesses close as purchasing power disappears. Municipal budgets shrink as tax bases erode. Younger generations leave in search of opportunity, often permanently. What remains is an aging population tied to places that no longer serve as engines of growth. Work had once been a source of pride and belonging. When the factory disappears, so does the social structure built around it.

These economic shifts translate directly into political change. Regions excluded from the gains of digital growth often view national institutions as distant or indifferent. Populist movements emerge from this resentment, fueled by the perception that prosperity is concentrated in places and among people who no longer share common experiences.

Technological displacement becomes cultural alienation. Cultural alienation becomes political realignment. Economic geography reshapes political identity.

A new spatial divide captures this transformation. The industrial Rust Belt contrasts sharply with what has come to be described as the *Cloud Belt*, a network of cities such as Seattle, San Francisco, Austin, Boston, and similar hubs across Europe and Asia. These regions attract disproportionate flows of capital, talent, and innovation. High wages, startup density, and global connectivity reinforce their dominance. Meanwhile, former manufacturing centers struggle to reinvent themselves in an economy that rewards digital fluency over physical output.

Generational migration accelerates this divergence. Younger workers gravitate toward technology centers that promise status, income, and cultural relevance. Their departure leaves behind communities with fewer resources and diminishing prospects. Social problems intensify as economic opportunity contracts. Meanwhile, digital hubs project confidence and optimism, reinforced by continuous innovation and proximity to global networks. Two economic cultures evolve side by side, separated not only by income but by worldview.

The contrast between abandoned factories and digital empires reveals a central truth of the *Techonomic* era. Power no longer flows primarily from material production. It flows from connectivity, information, and algorithmic scale. Physical infrastructure defined the last century. Data infrastructure defines the next. As the centers of economic gravity shift from land and machinery to code and cloud systems, entire societies are forced to confront a reordering of value that extends far beyond markets alone.

How Power Structures Are Being Rewritten

The rise of digital empires has fundamentally altered the architecture of global power. Authority that once flowed through parliaments, regulatory agencies, and corporate boards now moves quietly through algorithms, data pipelines, and platform infrastructure. Influence is no longer anchored to territory or traditional capital. It increasingly rests on the ability to shape information flows, guide user behavior, and coordinate networks at planetary scale. This gradual transfer of authority has produced what scholars describe as *algorithmic governance drift*, a condition in which private platforms exercise governance-like power without formal political mandate.

Control over information, finance, and communication has concentrated within a small constellation of technology firms. Recommendation systems determine which stories reach billions of screens. Payment networks decide which transactions are permitted or blocked. AI models influence hiring, lending, advertising, and political discourse. These mechanisms operate across borders, often beyond the effective reach of national regulators. Institutions designed for territorial oversight struggle to govern systems that function independently of geography.

The nature of authority itself has shifted. Industrial magnates once exerted power through factories, natural resources, and labor forces. Technocratic power arises from cloud platforms, application interfaces, and data governance. The operators of these systems do not merely compete within markets. They define the conditions under which markets operate. A platform update can alter the fortunes of thousands of businesses. A change in ranking logic can elevate or silence entire communities. A modification to payment infrastructure can determine who participates in the digital economy at all.

Governments increasingly find themselves negotiating with corporations that resemble geopolitical actors. Cloud providers host critical infrastructure for communication, defense, and public administration. Social media platforms influence electoral discourse. Cryptographic networks process cross-border value flows outside traditional banking oversight. States cannot simply impose authority through regulation alone, because the technologies in question have become integral to national functioning. Cooperation and conflict now coexist in a fragile symbiosis between governments and platforms.

Corporate alliances accelerate this transformation. Industry consortia establish standards, protocols, and interoperability frameworks that shape global markets long before legislation appears. These technical agreements function as informal rulemaking systems, determining how information moves, how payments settle, and how digital identity is verified. Because entire industries depend on them, such standards often become de facto law. Power shifts away from public deliberation and toward technical coordination, where engineers and executives quietly design the rules that govern daily life.

As oversight lags behind innovation, power becomes less visible and more diffuse. Machine learning systems automate decisions across finance, healthcare, commerce, and media without clear human attribution. Millions of micro-decisions accumulate into large-scale influence, shaping opportunity and behavior without public scrutiny. Democratic accountability weakens when authority is embedded in code rather than embodied in identifiable institutions. Responsibility fragments across layers of software, data, and optimization logic that few can inspect and no single actor fully controls.

Power in the *Techonomic* era flows through networks rather than institutions. Influence arises from connectivity, algorithmic leverage, and the ability to guide choices through system design rather than decree. As traditional hierarchies yield to digital sovereignty, societies confront a new reality in which technocrats and platform architects function as parallel governors over economic and social life. The central question is no longer only who owns wealth, but who controls the systems upon which modern civilization depends.

The Decline of Traditional Wealth Models

The foundations of traditional wealth, real estate, industrial capital, and physical commodities, are losing their primacy in a world shaped by digital networks and algorithmic infrastructure. Influence that once depended on land, factories, or shipping fleets is increasingly overshadowed by assets that scale without physical limits. This transition marks a period of *asset decoupling*, in which the relationship between tangible property and economic dominance breaks apart. Wealth no longer depends primarily on what one owns in the physical world, but on what one controls in the digital one.

Industrial-era capital was built on scarcity. A factory could produce only what its machinery and workforce allowed. A mine yielded only what it contained. Ports and railroads expanded influence, yet always within geographic boundaries. Digital wealth follows different rules. Platforms scale globally with minimal marginal cost. Network effects amplify influence as each new user adds value. Data accumulates as a renewable asset that grows more powerful with every interaction. Traditional capital struggles to compete with leverage of this kind.

This shift explains the rapid decline of firms that resist digital transformation. Kodak failed to adapt to digital photography. Sears could not reorient itself around ecommerce. Legacy media lost ground to streaming platforms capable of instantaneous global distribution. Even long-established banks face mounting pressure from fintech firms that use automation, blockchain, and AI to reduce costs and expand reach. Advantage that once protected incumbents now often becomes a liability when digital competitors move faster than institutions can respond.

A deeper inversion is underway. Intellectual property, cloud computation, and data orchestration have become the new pillars of economic power. Control over recommendation systems, digital identity frameworks, and platform ecosystems shapes commerce and communication at global scale. Wealth flows less from producing goods than from directing the movement of value across digital networks. Companies such as Amazon, Google, and Tencent dominate not because they own factories, but because they organize information, attention, and transactions with extraordinary efficiency.

Traditional wealth models also struggle because they depend on linear growth in an era defined by exponential scaling. Industrial output increases through incremental expansion. Digital platforms increase value through viral adoption, automation, and interconnected services. A single software update can activate millions of users at once. The asymmetry between physical and digital growth widens each year, leaving traditional firms structurally unable to match the speed or reach of digital systems.

These changes reshape global hierarchies. Land once defined aristocracy. Industrial capital defined the twentieth century elite. In the *Techonomic* era, power concentrates among those who control cloud platforms, algorithmic governance, AI infrastructure, and the invisible networks that mediate communication and commerce. Firms that master this architecture multiply their influence regardless of geography, while those anchored to material assets face long-term decline.

This transition is neither purely beneficial nor purely destructive. Digital wealth accelerates innovation, lowers barriers to entry, and creates opportunities for individuals without access to traditional capital. At the same time, it concentrates power within small groups that manage the infrastructure of modern life, raising concerns about accountability, resilience, and long-term stability. The attributes that enable rapid growth also enable rapid dominance.

Ultimately, traditional wealth models are losing relevance in an economy where power flows through connections rather than commodities. Factories, mines, and real estate still matter, but they no longer define the highest echelon of influence. Dominance in the *Techonomic* era belongs to those who shape digital systems, govern data, and orchestrate information at scale. This shift represents a profound realignment of wealth, opportunity, and power in the twenty-first century.

The Rise of Tech Billionaires

The *Techonomic* era has produced a new class of billionaires whose fortunes grow at speeds impossible in the industrial age. Unlike steel magnates or oil barons, whose wealth accumulated over decades, tech founders can ascend to global influence within a few years through the exponential mechanics of digital markets. This reflects what can be described as *hyperaccumulative wealth dynamics*, a pattern in which network effects, automation, and platform scalability amplify returns far faster than traditional economic systems ever allowed. Wealth no longer grows linearly. It compounds through code, connectivity, and adoption.

Modern examples illustrate this acceleration clearly. Elon Musk directs innovation across electric vehicles, aerospace, energy, and artificial intelligence, converting engineering vision into global leverage. Jeff Bezos transformed Amazon from an online bookstore into a logistics and cloud-computing backbone that supports much of the modern economy. Mark Zuckerberg governs communication platforms that shape discourse and culture for billions. Vitalik Buterin reimagined financial systems through programmable money and decentralized protocols. These figures do not merely build products. They construct architectures upon which industries, markets, and daily life increasingly depend.

The speed of ascent is matched by the speed of collapse in a system defined by digital volatility. The rise and fall of Sam Bankman-Fried offers a cautionary example. His exchange, FTX, became a major global financial platform in a remarkably short time. Venture capital firms, celebrities, and policymakers treated him as a bridge between crypto innovation and regulatory legitimacy. Investigations later revealed that customer funds were misused to cover losses at Alameda Research and to support political influence efforts. When confidence evaporated, billions in value disappeared almost overnight. Hundreds of thousands of users suffered losses. FTX became a case study in how rapid expansion without governance magnifies both risk and damage across interconnected markets.

What distinguishes tech billionaires from industrial magnates is not only the scale of their wealth, but the nature of the power it confers. Industrial tycoons controlled railroads, oil fields, and manufacturing plants. Modern technocrats control networks, data flows, digital identity, and the infrastructures that mediate everyday life. A change in an algorithm can alter public opinion. A policy adjustment within a cloud platform can affect millions of businesses. A feature update on a social network can reshape how societies communicate. Economic power merges with civilizational influence.

This transformation blurs the boundary between private enterprise and public governance. Tech billionaires increasingly occupy roles resembling cultural leaders, geopolitical actors, and custodians of critical infrastructure. Their platforms influence education, financial access, political narratives, and social norms. Governments negotiate with them not merely as corporations, but as operators of systems essential to national functioning. Infrastructure becomes inseparable from authority.

The consequences are mixed. On one hand, these innovators accelerate breakthroughs in medicine, transportation, energy, and communication that improve quality of life. They expand access to information, lower entry barriers for entrepreneurs, and enable global coordination at unprecedented scale. On the other hand, concentrated digital power challenges democratic oversight, intensifies inequality, and weakens public institutions. The same platforms that empower creativity also shape behavior, influence elections, and mediate human relationships beyond the reach of traditional accountability.

Modern societies now confront a novel reality. A small number of individuals can sway global markets, direct technological trajectories, and affect the daily lives of billions. Economic success becomes inseparable from control over digital infrastructure. Knowledge becomes the primary currency of influence. Wealth follows architecture rather than acreage.

Ultimately, tech billionaires embody the shift from industrial to algorithmic dominance. Their rapid ascent illustrates how value creation has migrated from physical production to platforms, data, and code. Their power reveals both the opportunity and the danger of a world in which innovation outpaces governance and influence scales faster than institutions can respond. This marks a new epoch in economic history, one in which mastery of digital systems determines who holds the levers of global power.

Chapter XIV

Industrial Insights

Historical Parallels with Industrial Revolution

The Industrial Revolution of the eighteenth and nineteenth centuries marked the first large-scale fusion of technology, labor, and capital. Mechanization transformed production, steam engines redefined power and mobility, and factories reorganized work around discipline, schedules, and centralized control. Agrarian hierarchies that had structured society for centuries began to collapse as industrialists displaced landowners. Entire populations migrated into expanding cities built around coal, steel, and machinery, reshaping both economic life and social order.

A recurring pattern in history emerges through what may be called *techno-historical mirroring*, a process in which one faction adopts disruptive technology or economic organization while its rival remains anchored to older systems. The result is not merely tactical advantage, but long-term civilizational realignment. Power shifts toward those who align themselves with emerging modes of production, coordination, and control.

Several conflicts illustrate this pattern with clarity.

During the American Civil War (A.D. 1861–1865), the Union embraced industrial warfare. It mobilized factories, railroads, telegraph networks, repeating rifles, and ironclad ships to sustain prolonged conflict. The Confederacy, by contrast, relied on agrarian production, decentralized logistics, and limited industrial capacity. The Union's ability to manufacture weapons at scale, move troops rapidly, and coordinate strategy through modern communication proved decisive. Victory followed not only from battlefield success, but from industrial sustainability.

Earlier examples reveal the same structural logic beneath conflicts often remembered primarily in religious terms. During the Pilgrimage of Grace in A.D. 1536, communities in northern England resisted Henry VIII's dissolution of the monasteries. While frequently framed as a theological uprising, the movement also reflected opposition to the rapid concentration of wealth and authority under the Tudor crown. Many participants viewed the seizure of monastic lands as both an economic dispossession and a religious rupture, interpreting the king's break with Rome as a departure from established moral and institutional order. The Tudor state, by contrast, redirected ecclesiastical property toward emerging commercial interests, strengthened centralized administration, and aligned itself with evolving property regimes. The resistance drew support from social structures rooted in a medieval agrarian system in which religious institutions functioned as centers of local welfare, land stewardship, and communal stability. The monarchy ultimately prevailed not through doctrinal resolution, but because centralized authority and proto-industrial organization proved more scalable than inherited religious and local arrangements.

The English Civil War (A.D. 1642–1651) followed a similar trajectory. Parliamentarian forces drew support from commercial classes and early industrial centers. They adopted disciplined military organization and leveraged the economic advantages of urban trade networks. Royalist forces depended more heavily on feudal loyalty, aristocratic patronage, and traditional cavalry warfare. Parliament's victory reinforced the rise of a commercial economy over a feudal one and laid institutional foundations for England's later industrial dominance.

These conflicts demonstrate that innovation is never neutral. New technologies and economic systems redistribute power, privilege those who adapt early, and erode the authority of those who resist. As mechanization spread during the nineteenth century, industrial capital eclipsed landed aristocracy. Steam power and rail networks redefined logistics, warfare, and political control, consolidating influence in the hands of those who mastered industrial organization.

The parallels to the *Techonomic* era are unmistakable. Where industrialization displaced artisanal labor and agrarian structures, artificial intelligence, blockchain coordination, and autonomous systems now reshape knowledge work, finance, and governance. The difference lies in speed. Transformations that once unfolded over generations now propagate globally within years through digital networks.

The historical lesson is clear. When disruptive technology emerges, those who adopt it gain structural advantage. Those who delay or resist often lose economic, political, and military influence. The Industrial Revolution demonstrated this across nations and centuries. The *Techonomic* era repeats the pattern at digital speed, reorganizing society around machine intelligence rather than mechanical force.

Economies of Scale and Cultural Shifts

The Industrial Revolution demonstrated that economies of scale do more than increase output. They compress culture itself. As production shifted into factories and mechanized systems, society reorganized around efficiency, uniformity, and predictability. Goods became standardized. Labor synchronized around shifts and schedules. Time itself was disciplined by clocks rather than seasons. Economic efficiency reshaped not only markets, but daily habits, expectations, and values.

As factories expanded, cost advantages favored large producers over small craftsmen. Mass production reduced prices and increased material access for many, but it also displaced artisanal identity and local independence. What emerged was a process that can be described as *cultural compression*, the narrowing of diverse ways of life into standardized patterns aligned with industrial efficiency. Regional customs gave way to national markets. Local rhythms yielded to uniform wages, standardized goods, and centralized planning.

These economic forces transformed family structures. Agrarian households once combined labor, childcare, and multigenerational cooperation within the home. Industrialization relocated work into factories, separating economic production from domestic life. The nuclear family replaced extended family networks. Gender roles shifted as wage labor redefined value around formal employment rather than household contribution. Industrial economies did not merely create jobs. They reorganized social life itself.

Technological infrastructure quietly reinforced these changes. Sanitation systems, electrification, public transportation, and climate control expanded the physical spaces in which people could live and work. Women entered public and economic life not only through legal reform, but because technology altered the conditions of participation. Mobility, hygiene, and mechanized household tools reshaped expectations around labor, education, and independence. Culture shifted because older arrangements became inefficient under new technological conditions.

The digital economy now produces a second wave of cultural compression, though through different mechanisms. Instead of factories enforcing uniform schedules, platforms enforce uniform interfaces. Instead of standardized physical goods, standardized digital experiences dominate daily life. Algorithms reward behaviors that scale efficiently, such as constant engagement, rapid consumption, and predictable interaction patterns. Cultural norms increasingly emerge from platform incentives rather than local tradition.

At first glance, digital systems appear liberating. Remote work dissolves geographic constraints. Online platforms enable global collaboration. Individuals gain access to knowledge, markets, and communities once restricted by location or class. Yet these same systems compress behavior by nudging users toward optimized patterns that favor attention capture, data extraction, and algorithmic predictability. Expression becomes shaped by what performs well within digital systems rather than by local meaning or communal continuity.

Family structures are again being reorganized. Remote labor blurs the boundary between home and work. Gig economies fragment stable employment into task-based income. Digital entrepreneurship enables earnings without institutional affiliation. These shifts offer flexibility and autonomy, but they also introduce instability, atomization, and weakened long-term social anchors. Economic incentives increasingly shape relationships, migration patterns, and life choices in subtle but persistent ways.

Both the industrial and digital eras reveal the same underlying dynamic. When economies scale, culture adapts. Efficiency reshapes norms. Systems reward behaviors that fit their architecture and marginalize those that do not. Material abundance and expanded access coexist with homogenization and social fragmentation.

Economies of scale do not merely change how societies produce wealth. They change how societies live. Industrialization compressed culture around factories and schedules. Digital systems compress culture around platforms and algorithms. Recognizing this pattern clarifies why technological progress consistently delivers both material gains and cultural loss, and why each new economic system quietly rewrites the terms of social life.

Why Scale Defeats Craft

Throughout economic history, human labor has been displaced not because it lacked value, but because it lost efficiency relative to machines. During the Industrial Revolution, skilled artisans produced high quality goods through time intensive manual processes. Their work carried craftsmanship, local identity, and pride. Yet mechanical looms, assembly lines, and steam powered factories could produce comparable goods faster, cheaper, and at far greater scale. Markets did not reward quality alone. They rewarded output, consistency, and price.

This dynamic follows what economists describe as cost displacement, the process by which human labor becomes economically uncompetitive when automation can deliver similar results at lower marginal cost. Machines do not tire, organize collectively, or demand wages. Once deployed, they compound productivity continuously. Even when human labor remains superior in nuance or quality, the price differential often becomes decisive. Consumers choose affordability. Firms choose scalability.

The same logic defines the *Techonomic* era. Artificial intelligence now replaces repetitive cognitive labor just as machines replaced repetitive physical labor. Tasks such as data entry, scheduling, customer service triage, basic legal review, and routine content generation can be performed faster and more consistently by algorithms. The issue is not that humans perform these tasks poorly. It is that automated systems perform them cheaply, continuously, and without interruption.

A useful comparison illustrates the shift. A craftsman producing a single item in a day may sell it directly to a customer, retaining most of the transaction's value. The exchange is personal, localized, and transparent. By contrast, an automated production system embedded within modern logistics platforms extracts value across many layers. Wages, taxes, corporate margins, platform fees, and financial intermediaries each claim a portion of the output. The individual worker receives only a fraction of the total value generated by the system.

This structure favors institutions aligned with scale. Corporations benefit from predictable output and continuous throughput. Governments benefit from payroll taxes, corporate taxes, and consumption taxes. Financial systems profit from transaction volume rather than craftsmanship. Automation integrates seamlessly into these structures, while individual labor does not. As a result, modern economies increasingly reward processes that maximize volume rather than those that preserve human participation.

Resistance to labor and economic displacement has appeared repeatedly throughout English history. During the transition from agrarian and artisanal economies to industrial capitalism, commoners resisted the enclosure of shared lands in the seventeenth and eighteenth centuries. Enclosure converted communal farmland into privately owned property optimized for large-scale production, often through Acts of Parliament that favored landowners, financiers, and state-aligned interests. Small farmers were displaced not because their labor lacked value, but because centralized ownership generated higher returns for elites and more predictable tax revenue for the Crown. What was lost was local autonomy and livelihood; what prevailed was efficiency, consolidation, and administrative control.

The same structural logic later appeared in the American colonies. During the Boston Tea Party of A.D. 1773, colonists resisted the imposition of the British East India Company, a state-backed monopoly whose losses were absorbed by Parliament while its market position was protected by law. What appeared as a protest over taxation was also a rejection of corporate privilege enforced by distant authority. In both cases, resistance was not directed against trade or innovation itself, but against the fusion of government power with centralized economic systems that displaced local producers and erased consent.

Modern automation accelerates this pattern. AI driven logistics systems replace dispatchers. Algorithmic trading replaces floor traders. Automated content systems displace entry level writers and editors. Each transition removes human labor not because humans lack competence, but because machines scale more efficiently within existing economic frameworks.

There is a utopian promise embedded in this shift. Automation can free humans from repetitive labor, reduce costs, and increase access to goods and services. Productivity gains can raise living standards if distributed broadly. The dystopian risk lies in misalignment. When the gains of automation flow primarily to owners of capital rather than displaced workers, inequality intensifies and social instability follows.

Handmade labor loses not because it lacks worth, but because modern economic systems reward efficiency over meaning. Automation centralizes value, amplifies capital, and marginalizes labor that cannot scale. The Industrial Revolution revealed this pattern through machinery. *Techonomics* extends it through algorithms. Understanding cost displacement clarifies why automation spreads relentlessly and why societies must consciously decide how efficiency gains are shared rather than assuming markets will resolve the imbalance on their own.

The Role of Innovation in Power Shifts

Throughout history, prolonged conflict between civilizations has hinged less on belief or bravery than on innovation. Power shifts occur when one side adopts new tools, organizational systems, or economic structures while its opponent remains anchored to older forms. Victory follows not because one side is morally superior, but because it operates within a more advanced technological paradigm.

One of the clearest examples is the long struggle between Western Christendom and the Islamic Caliphates that followed the early Muslim conquests of the seventh and eighth centuries. Within a century of Muhammad's death in A.D. 632, Islamic armies had conquered vast Christian territories across the Middle East and North Africa, including Syria, Egypt, Palestine, Jordan, and much of modern-day Turkey. These regions had once been core centers of Christianity, hosting ancient patriarchates, monasteries, and theological schools. Their loss permanently altered the religious and geopolitical map of the world.

The Crusades emerged as a response to this expansion. Beginning in A.D. 1095, Western armies attempted to reclaim territories that had once belonged to Christendom. Despite moments of success, including the capture of Jerusalem in A.D. 1099, the broader outcome was strategic failure. The lands of Syria, Egypt, and Anatolia remained under Islamic rule. Military courage and religious conviction proved insufficient against entrenched systems of power, logistics, and regional dominance.

The balance began to shift centuries later, not through theology, but through technology. Western Civilization halted further Islamic expansion at key moments, most notably at the Battle of Lepanto in A.D. 1571, where a coalition of European naval powers broke Ottoman dominance in the Mediterranean, and at the Siege of Vienna in A.D. 1683, which marked the end of sustained Ottoman advance into Central Europe. These were not decisive reversals of empire, but they signaled an emerging asymmetry. The Christian West was entering a period of sustained innovation in shipbuilding, artillery, finance, manufacturing, and scientific organization that would gradually compound into structural advantage.

By the late nineteenth century, the Ottoman Empire increasingly relied on Western-built ships, weapons, railways, and administrative models. Innovation had shifted outward. Dependency followed. Military courage and territorial scale could no longer compensate for lagging industrial capacity, logistics, and technological sovereignty. Power migrated to those who produced the systems of modernity rather than those who merely adopted them.

When the First World War erupted, this imbalance became decisive. The Ottoman Empire entered the conflict without independent industrial capacity sufficient to sustain modern warfare. Western powers, operating with mass production, mechanized logistics, and industrial scale coordination, dismantled Ottoman control with relative speed. By 1922, the empire collapsed, ending centuries of imperial dominance. The lesson was not ideological. It was structural. Civilizations that innovate internally preserve autonomy. Those that rely on external innovation become vulnerable.

This pattern repeats across military history. Gunpowder rendered swords obsolete. Firearms defeated armored knights. Artillery collapsed medieval fortifications. Tanks ended the dominance of horse cavalry. Aircraft rendered battleships secondary. In each case, the side that adopted the new technology reorganized warfare itself, while the opposing side attempted to fight the next war using the tools of the last one.

The *Techonomic* era introduces a comparable inflection point. The next major power shift is unlikely to resemble twentieth century warfare. One side may still industrially produce tanks, aircraft, missiles, and bombs at scale, relying on traditional military hardware and centralized command structures. The opposing side may instead deploy swarms of unmanned drones, autonomous vehicles, and AI coordinated systems that overwhelm conventional defenses through speed, quantity, and adaptive behavior.

Conflict will no longer be confined to physical battlefields. AI systems will attack electrical grids, water systems, logistics networks, and communications infrastructure. Power outages and supply disruptions will precede troop movement. Financial systems will become targets. Access to banking networks, payment rails, and settlement systems may be shut down remotely, freezing economic activity without firing a shot. Digital identity systems could be compromised or disabled, preventing citizens from accessing services, employment, or travel.

Information warfare will operate continuously. AI driven propaganda systems will inject tailored narratives directly onto personal devices, shaping public opinion on both sides of a conflict. Social media feeds, messaging platforms, and video networks will be flooded with synthetic content designed to confuse, divide, demoralize, or provoke. The objective will not be persuasion alone, but exhaustion and distrust. When populations cannot distinguish truth from fabrication, coordinated response collapses.

Diplomacy itself becomes vulnerable. Treaties, trade agreements, and international financial arrangements depend on digital trust and verification. A successful cyber offensive can undermine these systems silently, severing alliances and disrupting cooperation without formal declarations of war. Influence shifts toward those who control algorithms, data flows, and infrastructure rather than those who command the largest armies.

This represents a return to an old truth expressed through new tools. Innovation determines power. Those who master the emerging paradigm gain decisive advantage. Those who cling to legacy systems find themselves fighting with obsolete weapons against invisible adversaries. History shows that such transitions are rarely gradual. They arrive suddenly, often revealed only after the outcome is already decided.

In the *Techonomic* era, power flows to those who control artificial intelligence, digital infrastructure, and algorithmic coordination. Just as industrial capacity once defined global dominance, computational capacity now defines sovereignty. The lesson of history remains unchanged. When innovation shifts, power follows. Civilizations that fail to adapt do not merely lose battles. They lose the ability to shape the future.

What History Teaches about Resistance to Change

History shows with remarkable consistency that technological disruption is almost always met with resistance, and for understandable reasons. When new tools reorganize how value is created, they threaten not only livelihoods but identity, autonomy, and social order. Resistance is rarely rooted in ignorance alone. It is often a defense of human dignity against forces that appear indifferent to the costs of efficiency.

During the early nineteenth century, English textile workers known as the Luddites destroyed mechanized looms between A.D. 1811 and 1816. A loom is a machine used to weave threads into cloth. Before mechanization, weaving was a skilled craft practiced by individual artisans or families using hand looms. This work provided economic independence, local status, and generational continuity. The mechanized loom automated the process, allowing a single operator to produce cloth faster and cheaper than many skilled weavers combined. The loom did not merely improve productivity. It erased an entire economic structure.

Luddite resistance was not irrational hostility toward technology itself. It was a response to the loss of labor sovereignty, the ability of workers to control their craft, income, and pace of life. Mechanization centralized production, transferred power to factory owners, and transformed independent artisans into wage laborers subject to schedules, oversight, and market volatility. What was lost was not only employment, but agency.

This pattern repeats across eras. When technology reorganizes economic systems, those anchored to older structures experience displacement that extends beyond finances. Cultural meaning collapses alongside economic stability. Economists often describe this as *status quo bias*, the tendency to resist change, but this framing understates the depth of the disruption. Humans resist not because they fear novelty, but because they sense the erosion of purpose, skill, and self direction.

Technological asymmetry becomes even more decisive in warfare. History repeatedly demonstrates that armies equipped with superior tools, logistics, and coordination systems defeat those reliant on outdated methods. During the American Civil War, the Union leveraged industrial manufacturing, railroads, telegraphs, mass produced rifles, and centralized logistics. The Confederacy depended largely on agrarian production and decentralized supply chains. The outcome reflected not courage or conviction, but industrial capacity.

Earlier examples follow the same logic. The introduction of gunpowder weapons rendered medieval knights and stone fortifications obsolete. Cannon armed ships replaced boarding tactics and reshaped naval dominance. In World War I, machine guns, artillery, and industrial scale mobilization transformed combat into mechanized slaughter, favoring nations capable of sustained industrial output. In each case, technology altered the balance of power faster than cultural, ethical, or institutional systems could adapt.

This historical pattern challenges the modern assumption that technological progress is inherently benevolent. Technocrats often frame innovation as neutral or inevitable, yet history shows that technology amplifies power without regard for virtue. Mechanization increases efficiency, but it also centralizes control. Automation reduces labor, but it also removes agency. Surveillance improves security, but it erodes privacy. Systems optimized for output frequently reshape human behavior to serve the system rather than the person.

A deeper philosophical tension emerges. Is human labor meant to cultivate stewardship, responsibility, and moral formation, as emphasized in Christian Scripture and much of Western thought, where work is tied to dignity and purpose. Or is technology meant to eliminate labor entirely, transforming humans into passive consumers maintained by automated systems. When machines perform work, make decisions, and monitor behavior, efficiency increases, but independence declines.

Resistance to technological change, then, is not merely fear of progress. It is often a defense of human scale, local control, and moral agency. The danger lies not in innovation itself, but in unrestrained adoption without ethical boundaries. Societies that treat technology as an unquestionable good risk trading freedom for convenience and sovereignty for optimization.

Despite resistance, innovation almost always prevails. Mechanization reshaped economies despite protests. Industrial warfare dominated despite moral horror. Artificial intelligence and automation continue expanding today because economic incentives favor speed, scale, and control. Resistance may slow adoption, but it rarely reverses it. The decisive question is not whether technology will advance, but whether humans will retain authority over the systems they create.

History suggests a path forward grounded in preparation with restraint. Denial magnifies disruption, but blind acceptance magnifies harm. Policymakers, educators, and citizens must evaluate not only what technology can do, but what it should do. Societies that integrate innovation while preserving human dignity, privacy, and responsibility remain more stable than those that sacrifice these values for efficiency alone.

Ultimately, history teaches that technological progress determines outcomes in war, economics, and power, but it does not determine meaning. Survival belongs to those who adapt wisely rather than blindly. Innovation will reshape the world whether embraced or resisted. The enduring challenge is whether humanity guides that transformation in service of human flourishing, or allows efficiency and control to replace freedom, labor, and moral responsibility altogether.

Chapter XV

Next Horizons

Predictions for the Next Decade

The next decade will unfold beneath a new form of geopolitical struggle commonly described as *Fifth Generation Warfare*. Earlier generations of conflict were defined by physical force and industrial capacity. First generation warfare relied on massed infantry. Second generation conflict emphasized gunpowder, artillery, and naval projection. Third generation warfare introduced mechanized tanks and aircraft. Fourth generation conflict shifted toward decentralized insurgencies and asymmetric tactics. Fifth generation conflict departs from these models entirely. Its primary theater is no longer territory, but perception. Power is exercised through information, influence, and psychological control rather than direct kinetic dominance.

As with every major technological transformation since the Industrial Revolution, emerging innovations migrate rapidly into military and state systems. Steam engines, railroads, and mechanized factories enabled industrial empires and total war. Today, artificial intelligence, autonomous systems, cyber capabilities, and predictive surveillance are being absorbed into modern defense doctrines. These tools expand state capacity while also increasing the risk of escalation, miscalculation, and instability. A technological arms race is underway once again, not for land or resources, but for control over digital systems and informational space.

This shift is already visible in contemporary conflicts. AI guided drones, autonomous targeting platforms, and coordinated swarms capable of operating with limited human oversight have appeared in Middle Eastern and Eastern European theaters. Cyber operations target power grids, communication networks, and financial infrastructure. Bot networks conduct psychological operations by flooding digital environments with competing narratives designed to confuse memory, manipulate emotion, and fracture social cohesion. Warfare increasingly proceeds without formal declarations or visible front lines. Conflict becomes continuous, diffuse, and embedded within civilian life, where the objective is not decisive victory but sustained influence.

Narrative control sits at the center of this landscape. Fifth generation conflict prioritizes the management of information through censorship, deplatforming, and algorithmic suppression of dissent. Labels such as misinformation, extremism, or foreign interference are used to marginalize viewpoints that fall outside approved frameworks. Individuals who diverge from dominant narratives risk digital exclusion rather than physical repression. The strategic goal is not conquest, but consent. Populations are shaped through filtering, curating, and directing what they are permitted to see, discuss, or believe.

At the same time, everyday life continues to reorganize under technological pressure. Education is likely to fragment further as artificial intelligence enables parents and individuals to generate customized curricula, assessments, and interactive learning environments with minimal institutional involvement. Podcasts, audiobooks, and informal online academies increasingly compete with traditional schooling as AI tutors improve in sophistication. The irony is sharp. During the pandemic, many assumed the remote office represented a permanent future. Only a few years later, many of the same white-collar roles that moved online face automation. Within a single decade, work itself has been redefined twice.

Cultural divisions are therefore likely to deepen. Those who embrace artificial intelligence will gain productivity, income, and influence. Those who resist it will often do so on moral, artistic, or existential grounds. Movements opposing AI in film, music, writing, and creative industries have already emerged. Some argue that authenticity must remain human. Others adopt machine tools aggressively, widening ideological and economic distance between acceleration and preservation. This divide mirrors earlier cultural fractures produced by industrialization, when mechanization reshaped labor faster than society could reconcile its meaning.

The coming decade thus represents not merely an era of innovation, but a civilizational transition comparable to the upheavals of the nineteenth century that overturned monarchies and reordered global power. Change will proceed rapidly. In 2020, few major corporations invested seriously in artificial intelligence or crypto infrastructure. By the mid 2020s, the world's most valuable firms are structured around both. Identity, governance, warfare, and culture are being rewritten within the span of a single generation.

The horizon ahead is not defined by new tools alone. It marks a reordering of authority and daily experience. Technology no longer automates labor in isolation. It increasingly automates perception, belief, and social coordination. Over the next ten years, civilizations will compete less over industrial output and more over informational sovereignty. This is the defining condition of *Techonomics*. Technology becomes the battlefield. Data becomes the weapon. The human mind becomes the contested territory.

Corporate Ecosystems and Neo-Industrial Models

The next phase of economic organization is no longer defined by isolated firms competing within narrow industries. It is defined by interconnected corporate ecosystems that operate less like traditional companies and more like digital states. In the *Techonomic* era, dominant enterprises build platforms, partnerships, data pipelines, and integration layers that extend influence far beyond any single product or service. Amazon does not merely sell goods. It coordinates logistics, retail marketplaces, cloud infrastructure, and artificial intelligence. Tencent integrates communication, payments, entertainment, and cloud services into a unified economic environment. In these models, value arises from orchestration rather than production alone.

This shift marks a neo-industrial phase that combines industrial scale with digital intelligence. Unlike legacy manufacturers constrained by factories, fixed labor, and geographic footprint, ecosystem firms scale through software rather than construction. Expansion occurs through code. Integration replaces physical accumulation. Entire sectors can be reshaped through platform design rather than capital investment in plants or equipment. The digital platform functions as a factory of coordination, producing value algorithmically rather than mechanically.

A defining feature of these ecosystems is that they operate as environments rather than enterprises. Traditional firms competed by outperforming rivals within defined markets. Ecosystem firms compete by absorbing functions that once required multiple institutions.

Communication, payments, logistics, media distribution, identity verification, and data analytics converge inside a single infrastructure. As these environments expand, they become gateways to economic participation. Smaller firms increasingly survive not by competing with ecosystems, but by operating within them.

Network effects intensify this dominance. Each additional participant increases the value of the system, attracting more users, more data, and more institutional dependence. Market power no longer rests primarily on supply chains or physical scale, but on integration depth and informational advantage. Firms that fail to achieve ecosystem status face structural dependency, marginalization, or acquisition rather than direct competition.

Competition itself is reshaped. Strategic alliances, shared platforms, and interoperable data environments increasingly replace traditional rivalry. Coordination across institutional boundaries becomes a necessity rather than a strategy. The neo-industrial model rewards systems that can integrate finance, technology, and logistics into a seamless whole. Fragmented firms struggle to match the efficiency and reach of coordinated networks.

This architecture introduces new forms of fragility. When a single platform spans commerce, communication, finance, and infrastructure, failure in one domain can cascade across the entire system. The risk resembles global finance prior to the 2008 crisis, but with deeper penetration into daily life. Platform concentration creates systemic vulnerabilities that existing governance frameworks are ill-equipped to manage. Resilience becomes as important as scale, yet incentives still favor expansion over redundancy.

Despite these risks, the trajectory is unmistakable. Coordinated ecosystems are replacing isolated enterprises just as factories once replaced workshops. Corporate networks now organize economic life in the way industrial giants once organized cities. Those who master orchestration rather than production will define the next era of wealth, influence, and institutional power.

The Future of Governance and Currency

Traditional systems of governance are entering a period of strain unseen since the early industrial age. National legislatures, constitutional frameworks, and regulatory institutions were designed for a world in which economic change unfolded slowly, information moved with friction, and authority flowed clearly from state to citizen. In the *Techonomic* era, those assumptions no longer hold. Artificial intelligence, decentralized networks, and digital currencies evolve faster than legislative cycles, judicial review, or electoral mandates. Governance increasingly reacts to technology rather than directing it.

Digital currency represents one of the clearest fault lines in this transformation. Cryptocurrencies, tokenized assets, and central bank digital currencies challenge the foundations of monetary sovereignty. Money is no longer merely a medium of exchange or a store of value. It becomes programmable, conditional, and traceable. Transactions can be automated, restricted, or reversed according to predefined rules embedded in code. These capabilities promise efficiency and inclusion, yet they also introduce unprecedented capacity for behavioral control within economic life.

Blockchain infrastructure further complicates governance. Distributed ledgers reduce reliance on traditional intermediaries and introduce transparency into financial and administrative systems. Smart contracts automate obligations that once depended on trust, negotiation, or institutional enforcement. Fraud can be reduced, settlement accelerated, and compliance automated. At the same time, authority migrates from human discretion toward technical architecture. Rules that were once debated publicly become embedded in systems that execute automatically and without appeal.

As a result, governance increasingly operates in hybrid form. States and corporations collaborate on standards, enforcement mechanisms, and digital infrastructure. In many cases, private platforms define operational rules first, and governments attempt regulation afterward. Content moderation, payment access, digital identity, and even political discourse increasingly occur inside privately owned systems. Authority becomes layered, fragmented, and less visible. Citizens experience governance through interfaces rather than institutions, often without clarity regarding who sets the rules or how they may be challenged.

History offers a cautionary parallel. Political systems have repeatedly struggled to absorb technological revolutions without disruption. The constitutional frameworks of the eighteenth century were reshaped by industrialization, mechanized warfare, and global trade. Likewise, postwar political institutions are now being reshaped by digital communication, algorithmic decision systems, and platform governance. What once required legislation and public deliberation increasingly occurs through software updates, protocol changes, and revised terms of service.

The geopolitical consequences are substantial. Nations that successfully integrate digital currency systems, AI oversight, and technological governance gain influence over global standards and trade frameworks. Those that resist or delay risk exclusion from financial networks, supply chains, and innovation ecosystems. At the same time, corporations capable of defining technical protocols and infrastructure accumulate authority rivaling that of states. Control over payment rails, identity systems, and cloud infrastructure becomes a form of geopolitical leverage.

The future of governance therefore depends on whether societies can reclaim agency over the systems they rely upon. An optimistic path includes transparent oversight, plural participation, and accountable design embedded into digital architecture from the outset. A darker path leads toward centralized digital control, erosion of sovereignty, and coercive dependency masked as efficiency. Governance will not disappear, but it will be redefined by how power flows through networks rather than borders.

Ethical and Philosophical Questions Ahead

The rise of *Techonomics* forces civilization to confront ethical questions that cannot be resolved through efficiency alone. As artificial intelligence and digital systems increasingly shape decisions in finance, employment, healthcare, and security, responsibility becomes difficult to locate. When an algorithm denies credit, flags a citizen, or influences a legal outcome, accountability fragments. The developer points to data. The platform points to automation. The institution points to compliance. Moral responsibility dissolves into systems that no single individual fully controls.

Inequality deepens under these conditions. Wealth increasingly concentrates among those who own platforms, data, and infrastructure rather than those who perform labor. Surveillance expands in parallel. Governments and corporations justify continuous monitoring in the name of safety, convenience, or optimization. The core ethical question is whether freedom can survive constant observation. Privacy shifts from a natural right to a conditional privilege, preserved only by those willing to withdraw from modern economic life.

Participation itself becomes compulsory. Digital systems track location, transactions, behavior, and biometric signals. Homes, vehicles, workplaces, and public spaces are networked by default. Refusal is rarely practical. Economic and social participation increasingly requires visibility. Agency erodes not through overt coercion, but through dependency on systems that cannot be meaningfully declined without exclusion.

At a deeper level, *Techonomics* challenges assumptions about human purpose. Are human beings meant to labor, create, and exercise stewardship over their lives, or are they meant to be optimized out of production altogether. If machines outperform humans in productivity, does labor retain moral value. If algorithms determine outcomes more efficiently than people, does judgment still matter. These are not engineering problems. They are philosophical ones that strike at the foundations of dignity and responsibility.

The meaning of work is destabilized. Identity has long been bound to contribution and effort. When automation replaces labor, dignity risks becoming detached from participation in productive life. Governance faces a parallel dilemma. When decision making becomes algorithmic, consent becomes indirect and authority opaque. Citizens may comply with systems they neither understand nor have the ability to challenge.

The ethics of wealth creation further complicate the picture. Vast fortunes now emerge from platforms that extract attention, data, and behavior rather than from visible labor or material production. This raises a fundamental question of legitimacy. Can democratic oversight meaningfully constrain systems that operate across borders, beyond jurisdiction, and at machine speed. When value is generated by architecture rather than effort, traditional moral frameworks strain to explain who deserves what and why.

Addressing these dilemmas requires intentional design rather than retroactive justification. Ethics cannot remain an afterthought appended to innovation. Moral reasoning must shape technological deployment, governance structures, and education itself. Without this integration, societies risk constructing systems that function with precision while hollowing out human dignity.

Techonomics therefore presents civilization with a choice. Technology can be ordered toward human flourishing, or humanity can be reordered to serve technological systems. The outcome depends not on innovation alone, but on whether moral capacity evolves alongside technical power.

Preparing for the Unknown

The defining condition of the *Techonomic* era is uncertainty. Change is rapid, interconnected, and nonlinear. For individuals, survival and advancement depend less on specialization than on adaptability. Technical proficiency remains important, but it is insufficient without critical reasoning, ethical judgment, and emotional resilience. Lifelong learning is no longer an aspiration. It is the minimum requirement for stability in a system that continually reshapes itself.

Organizations face the same reality. Strategies built around single products, static markets, or rigid hierarchies fracture under conditions defined by platforms, automation, and continuous disruption. Efficiency, once the supreme advantage, increasingly gives way to resilience. Firms that endure are those that cultivate experimentation, cross-disciplinary thinking, and the ability to pivot without institutional collapse. Culture becomes as consequential as capital.

Societies, too, must abandon reactive governance. Policies designed to address disruption after it occurs arrive too late to prevent damage. Investment in education, retraining, and digital access must be treated as foundational infrastructure rather than supplemental support. Communities that prepare for transition preserve cohesion. Those that cling to industrial assumptions experience fragmentation as change compounds faster than institutions can absorb it.

Foresight becomes a strategic discipline. Scenario planning, predictive modeling, and continuous reassessment allow institutions to anticipate disruption rather than respond to crisis. Risk management shifts from compliance to anticipation. The ability to prepare for multiple futures becomes a source of strength rather than indecision.

Psychological preparation is equally essential. Stability is no longer the default condition of economic life. Volatility increasingly signals transformation rather than collapse. Individuals and institutions that interpret disruption as cyclical rather than catastrophic respond with flexibility instead of fear. In such environments, uncertainty becomes navigable rather than paralyzing.

Preparation is ultimately collective. Learning, governance, and innovation cannot be separated. Individuals who understand systems rather than isolated tasks, organizations that align technological progress with human needs, and governments that design adaptive frameworks together shape resilient societies. Fragmented responses magnify instability. Coordinated adaptation absorbs it.

In this sense, preparedness is not defensive. It is creative. It determines whether *Techonomics* becomes a force that expands human agency or a system that overwhelms it. The future will not reward those who predict it perfectly, but those who remain capable of learning, judging, and acting wisely as it unfolds.

Ending Note

This book began with a simple claim: technology is no longer one industry among many. It has become the infrastructure of modern civilization. By the time we reach this final section, that truth should no longer feel theoretical. *Techonomics* now shapes how wealth is created, how authority is exercised, how information flows, and how ordinary people participate in economic and civic life. The question is no longer whether technology will influence society, but whether society can still influence the direction technology takes.

Throughout history, periods of rapid innovation have produced both extraordinary progress and profound instability. The Industrial Revolution expanded human capacity while uprooting communities, concentrating power, and triggering conflicts whose consequences lasted generations. The digital transformation now unfolding follows the same historical pattern, but at greater speed and broader scale. Where industry reorganized labor and land, *Techonomics* reorganizes perception, identity, and coordination itself. Power no longer depends primarily on ownership of physical assets. It depends on control of systems, data, and networks.

The central danger of this era is not collapse, but capture. Movements arise in response to injustice, stagnation, or exclusion. They promise decentralization, autonomy, and renewal. Yet history shows that systems rarely surrender power willingly. They adapt, absorb, and reassert control through complexity, regulation, and scale. The digital economy now reflects this pattern. Technologies once framed as tools of liberation increasingly function as instruments of administration. Participation remains possible, but only on terms set by systems few fully understand.

This does not mean technological progress is inherently destructive. Innovation has always been morally neutral in itself. It amplifies intention rather than replacing it. The same tools that enable surveillance can enable transparency. The same networks that centralize power can distribute knowledge. The same automation that displaces labor can free human energy for higher purposes. The outcome depends not on the tools, but on the values embedded within them and the courage of those who govern their use.

The deeper questions, therefore, are not economic alone. They are philosophical. What is human work for? Is labor merely an inefficiency to be eliminated, or is it a source of dignity, responsibility, and self sovereignty? Is society meant to optimize efficiency at any cost, or to cultivate human flourishing even when it is slower, messier, and less profitable? These questions cannot be answered by algorithms. They require moral judgment rooted in tradition, history, and an honest understanding of human nature.

Ending Note

If the Industrial Revolution built the cocoon, *Techonomics* represents what emerged from it. The infrastructure remains. The raw materials remain. But the purpose has shifted. Power today is exercised less through force than through access. Less through ownership than through permission. Less through visible authority than through systems that quietly determine what is possible. A society that does not understand this transformation risks mistaking convenience for freedom and access for sovereignty.

Yet the future is not closed. History does not move automatically toward tyranny or liberation. It moves according to the decisions of individuals, institutions, and cultures. Societies that cultivate literacy, discipline, and moral clarity retain agency. Those that surrender judgment in exchange for comfort do not. The same technologies that threaten autonomy can strengthen it if governed wisely. The same systems that concentrate power can be resisted through decentralization, education, and deliberate restraint.

Preparedness in this era means more than technical skill. It requires historical awareness, ethical grounding, and the willingness to remain human in systems designed to optimize away human complexity. It requires families, communities, and institutions capable of saying no as well as yes. It requires leaders who understand that not every innovation deserves adoption, and citizens who understand that convenience is never free.

Techonomics will continue to advance whether welcomed or resisted. That much history makes clear. What remains undecided is whether it will serve humanity or quietly subordinate it. The system will continue to evolve. The question is whether people will evolve with discernment, or drift into dependency without noticing when agency was surrendered.

This book does not offer certainty. It offers orientation. It traces patterns so they can be recognized. It names forces so they can be confronted. It does not ask the reader to reject technology, nor to worship it. It asks something far more demanding. To understand it.

The future will belong neither to those who deny technological change nor to those who embrace it without question. It will belong to those who retain judgment when systems grow powerful, who preserve human dignity when efficiency tempts surrender, and who remember that civilizations are not defined by what they invent, but by what they choose to protect.

Civilizations endure not by perfecting their machines, but by remembering what those machines must never be allowed to replace.

KEY TERMS AND CONCEPTS

01) **Advanced Research Projects Agency (ARPA):** A U.S. government research agency created in 1958 during the Cold War to ensure technological superiority. ARPA funded early work in computing, networking, and systems research that private industry was unwilling or unable to pursue. Its role illustrates how state investment shaped the foundations of the digital age.

02) **Advanced Research Projects Agency Network (ARPANET):** ARPANET was the first large-scale computer network to use packet switching, a method of sending information in small pieces across multiple paths. Designed for resilience rather than efficiency, it demonstrated that communication systems could survive disruptions. ARPANET eventually evolved into the modern internet.

03) **Agricultural telemetry:** Agricultural telemetry refers to the use of sensors, data collection, and analytics to monitor crops, soil conditions, equipment, and weather. These systems allow farmers to make decisions based on real-time data rather than experience alone. It reflects the expansion of digital infrastructure into biological and environmental systems.

04) **Artificial Intelligence (AI):** Artificial intelligence describes computer systems designed to perform tasks associated with human cognition, such as recognizing images, generating language, or making predictions. Modern AI relies heavily on statistical models trained on large datasets rather than explicit rules. Its power comes from scale, computation, and data availability.

05) **AI-connected tokens:** AI-connected tokens are digital assets whose value or function is tied to artificial intelligence systems or services. They may represent access to AI models, payment for computation, or governance rights within AI platforms. These tokens attempt to financialize AI infrastructure.

06) **AI Fine Tuning:** AI fine tuning is the process of adapting a general-purpose AI model to a specific task or domain using additional training data. This allows models to become more accurate and context-aware. Fine tuning trades broad flexibility for specialization.

07) **AI Inference:** Inference is the stage where an AI model produces outputs based on new input data. This is the phase users interact with directly, such as asking a chatbot a question or receiving a recommendation. While training is expensive and centralized, inference is often deployed widely.

08) **AI Overfitting:** Overfitting occurs when an AI model learns its training data too precisely and fails to generalize to new situations. This leads to confident but unreliable outputs. Overfitting highlights the limits of pattern recognition without true understanding.

09) **AI Training:** AI training involves feeding large datasets into models and adjusting their parameters through repeated computation. Training determines what patterns an AI system can recognize and reproduce. It requires significant energy, hardware, and data resources.

10) **Algorithms:** Algorithms are structured procedures that instruct computers how to process data or make decisions. In the digital economy, algorithms shape search results, prices, content visibility, and access. They increasingly function as invisible rule-makers.

11) **Algorithmic externalities:** Algorithmic externalities are unintended consequences produced by automated systems. These may include bias, feedback loops, social polarization, or economic distortion. Such effects often emerge only at scale.

12) **Algorithmic governance:** Algorithmic governance refers to the use of software systems to shape rules, enforcement, and decision-making processes. This can occur in finance, content moderation, or public administration. Authority shifts from human judgment to encoded logic.

13) **Algorithmic governance drift:** Governance drift occurs when algorithmic systems gradually diverge from their original purpose as conditions change. Models adapt faster than laws or institutions. This creates gaps between intent and outcome.

14) **Alternative data modeling:** Alternative data modeling uses nontraditional data sources such as transaction behavior, location data, or online activity to assess risk or value. It expands financial analysis beyond credit scores and balance sheets. This approach raises questions about privacy and consent.

15) **Altcoins:** Altcoins are cryptocurrencies other than Bitcoin. They experiment with different technologies, governance models, and economic incentives. Many attempt to improve speed, scalability, or programmability.

16) **Anticipatory governance:** Anticipatory governance aims to address technological risks before they fully emerge. It emphasizes foresight, scenario planning, and adaptive regulation. The goal is to reduce harm without halting innovation.

17) **Application Programming Interface (API):** An API is a standardized way for software systems to communicate with each other. APIs allow platforms to expose specific functions without revealing internal systems. They enable ecosystems, integrations, and modular innovation.

18) **Asymmetric cryptography:** Asymmetric cryptography uses a pair of keys: a public key for encryption and a private key for decryption. This system enables secure communication over open networks. It forms the basis of digital identity, blockchain security, and online trust.

19) **Asset decoupling:** Asset decoupling refers to the separation of economic value from physical objects or traditional institutions. Digital assets can exist independently of geography or centralized intermediaries. This shifts how ownership and trust are defined.

20) **Attack surface:** An attack surface is the total number of points where a system can be exploited. As systems become more complex and interconnected, attack surfaces expand. Security becomes harder to manage at scale.

21) **Attention economy:** The attention economy treats human focus as a scarce resource to be captured and monetized. Platforms compete to maximize engagement, often using behavioral data and algorithms. Attention becomes a form of currency.

22) **Automated decision systems (ADS):** ADS are systems that make or influence decisions without direct human involvement. They are used in credit approval, hiring, policing, and content moderation. Their opacity raises concerns about accountability.

23) **Automated market makers:** Automated market makers are algorithmic systems that enable trading without traditional order books. Prices are set by mathematical formulas based on available liquidity. They replace human intermediaries with code.

24) Avalanche (AVAX) Layer 1: Avalanche is a base-layer blockchain designed for speed and scalability. It uses novel consensus mechanisms to achieve fast transaction finality. The network supports customizable sub-networks.

25) Banking-as-a-service: Banking-as-a-service allows companies to offer financial products without being banks themselves. Licensed institutions provide the backend infrastructure through APIs. This model modularizes financial services.

26) Behavioral telemetry: Behavioral telemetry is the continuous collection of data about user actions, preferences, and habits. It enables prediction, personalization, and optimization. Over time, it becomes a detailed behavioral map.

27) Big data: Big data refers to extremely large datasets generated by digital activity, sensors, and platforms. Its value lies not in individual records but in patterns across populations. Big data powers AI and predictive systems.

28) Bitcoin (BTC): Bitcoin is a decentralized digital currency introduced in 2009. It enables peer-to-peer transactions without banks or governments. Bitcoin established scarcity and trust through cryptography and consensus.

29) Bitcoin (BTC) Layer 1: Bitcoin's base layer records transactions on a public, immutable ledger. It prioritizes security and decentralization over speed. Changes occur slowly to preserve stability.

30) Bitcoin (BTC) Layer 2: Layer 2 systems operate on top of Bitcoin to improve transaction speed and cost. They handle activity off the main chain while settling back to it. This introduces efficiency but adds complexity.

31) Bitcoin white paper: The Bitcoin white paper outlines a system for digital money without trusted intermediaries. It introduced blockchain, mining, and decentralized consensus. The document became foundational to digital assets.

32) Block explorers: Block explorers are tools that allow users to view blockchain data. They provide transparency into transactions, addresses, and network activity. Explorers make public ledgers readable.

33) Blockchains: A blockchain is a distributed ledger maintained by a network rather than a central authority. Once recorded, data cannot easily be altered. Blockchains enable shared trust without centralized control.

34) **Blocks:** Blocks are collections of transactions added sequentially to a blockchain. Each block references the previous one, creating an immutable chain. This structure ensures integrity.

35) **Bot:** A bot is an automated software agent that performs tasks without human input. Bots operate in markets, messaging systems, and infrastructure management. They can amplify efficiency or manipulation.

36) **Bot farms:** Bot farms are coordinated networks of automated accounts. They are often used to influence markets, social media, or political discourse. Scale magnifies their impact.

37) **Broadband internet:** Broadband provides high-speed, continuous internet access. It enabled streaming, cloud services, and digital platforms. Broadband transformed the internet from a niche tool into a utility.

38) **Buy Now Pay Later (BNPL):** BNPL allows consumers to split purchases into installments, often without interest. Credit becomes embedded directly into commerce. This shifts borrowing from banks to platforms.

39) **Cardano (ADA) Layer 1:** Cardano is a base-layer blockchain that emphasizes formal verification and academic research. Its design prioritizes security, sustainability, and long-term governance over rapid experimentation. Development proceeds slowly by design.

40) **Central Bank Digital Currencies (CBDCs):** CBDCs are digital forms of national currencies issued directly by central banks. They aim to modernize payment systems while preserving state control over money. Unlike cryptocurrencies, CBDCs are centralized and policy-driven.

41) **CERN:** The European Organization for Nuclear Research is a scientific research institution best known for particle physics. CERN played a pivotal role in digital history as the birthplace of the World Wide Web. It illustrates how scientific infrastructure can produce civilizational technologies.

42) **CHIPS and Science Act:** The European Organization for Nuclear Research is a scientific research institution best known for particle physics. CERN played a pivotal role in digital history as the birthplace of the World Wide Web. It illustrates how scientific infrastructure can produce civilizational technologies.

43) **Classical Economics:** An economic framework emphasizing land, labor, and capital as the primary drivers of value. Classical economics assumes scarcity, diminishing returns, and physical constraints. Many digital-era dynamics strain these assumptions.

44) **Cloud Belt:** A term describing geographic regions that host concentrated cloud computing infrastructure. These areas gain economic and strategic importance due to data gravity and energy access. The Cloud Belt reshapes geopolitical power.

45) **Cloud computing:** Cloud computing provides on-demand access to computing resources over the internet. It replaces local hardware with centralized data centers. This shift enables scale but increases dependency on a few providers.

46) **Cognitive inequality:** The growing gap between individuals or groups based on access to tools that enhance cognition, such as AI systems. Cognitive inequality compounds traditional economic inequality. It reshapes education, labor, and power.

47) **Colocation centers:** Facilities where multiple organizations house servers and networking equipment. Colocation reduces costs while maintaining physical control. These centers form the backbone of internet infrastructure.

48) **Computing:** Computing refers to the manipulation of information through machines. Over time, it has evolved from calculation to simulation, prediction, and optimization. Computing is now embedded in nearly all systems.

49) **Consensus mechanisms:** Processes used by distributed networks to agree on shared data. In blockchains, consensus replaces centralized authority. Different mechanisms trade off speed, security, and decentralization.

50) **Constitutional AI:** An approach to AI alignment where systems are guided by explicit principles rather than ad hoc rules. These principles act as a moral or operational framework. The goal is more predictable and constrained behavior.

Key Terms and Concepts

51) **Coordinated Inauthentic Behavior (CIB):** Organized efforts to manipulate discourse using fake or automated accounts. CIB undermines trust in digital communication. Platforms struggle to detect it at scale.

52) **Cross-chain tokens:** Tokens designed to operate across multiple blockchains. They aim to improve interoperability and liquidity. Cross-chain systems increase complexity and risk.

53) **Cryptocurrencies:** Digital currencies secured by cryptography and maintained by decentralized networks. They challenge traditional money by removing centralized intermediaries. Their volatility reflects unresolved questions of value and governance.

54) **Cryptographic keys:** Mathematical credentials that control access to digital assets or systems. Ownership is defined by possession of keys rather than identity. Key management becomes personal responsibility.

55) **Cryptography:** The practice of securing information through mathematical techniques. Cryptography enables privacy, authentication, and trust in digital systems. It underpins modern finance and communication.

56) **Cultural compression:** The rapid collapse of historical time and cultural distance due to digital media. Ideas, norms, and conflicts spread instantly across societies. This accelerates social change and friction.

57) **Cybernetic markets:** Markets increasingly governed by feedback loops between algorithms, data, and behavior. Prices respond to machine-driven signals rather than human judgment. These markets can be fast but fragile.

58) **Dark fiber:** Unused or underutilized fiber optic cables. Dark fiber represents latent capacity for future network expansion. Control over dark fiber is strategically valuable.

59) **Data:** Recorded information generated by human activity, machines, or natural systems. Data becomes economically valuable when aggregated and analyzed. It is the raw material of the digital economy.

60) **Data centers:** Facilities that house servers, storage, and networking equipment. Data centers concentrate computation, energy use, and geopolitical importance. They are physical anchors of the digital world.

61) **Data gravity:** The tendency for data to attract applications, services, and infrastructure. Large datasets are costly to move. Data gravity reinforces platform dominance.

62) **Data localization:** Policies requiring data to be stored within national borders. These laws reflect concerns about sovereignty, privacy, and control. Localization fragments the global internet.

63) **Data pipelines:** Systems that move data from collection to storage to analysis. Pipelines automate the flow of information across organizations. Their design determines what can be known and acted upon.

64) **Decentralized applications (dApps):** Applications that run on blockchain networks rather than centralized servers. They aim to reduce reliance on intermediaries. Usability and scalability remain challenges.

65) **Decentralized Autonomous Organization (DAO):** An organization governed by smart contracts and token-based voting. DAOs automate coordination and governance. They blur the line between software and institutions.

66) **Decentralized exchanges (DEXs):** Platforms that allow users to trade digital assets without centralized custody. Trades are executed by smart contracts. DEXs prioritize autonomy over convenience.

67) **Decentralized finance (DeFi):** A system of financial services built on blockchain infrastructure. DeFi recreates lending, trading, and derivatives without traditional intermediaries. It exposes both innovation and systemic risk.

68) **Decentralized identity systems:** Identity frameworks where individuals control their credentials rather than centralized authorities. These systems aim to reduce surveillance and data exploitation. Adoption remains limited.

69) **Decentralized oracles:** Systems that bring external data onto blockchains. Oracles bridge real-world information and on-chain logic. They are critical points of trust.

70) **Defense Advanced Research Projects Agency (DARPA):** The successor to ARPA and a major funder of advanced research. DARPA supports high-risk projects with military and civilian implications. Its influence spans computing, AI, and networking.

71) **Deepfakes:** Synthetic media generated using AI to mimic real people. Deepfakes blur the line between reality and fabrication. They challenge trust in visual evidence.

72) **Derivatives protocols:** Blockchain-based systems for trading financial derivatives. They automate leverage and risk through smart contracts. Errors can propagate rapidly.

73) **Dial-up internet:** Early internet access using telephone lines. Dial-up shaped the first online experiences. Its limitations influenced early web design.

74) **Digital asset gaming:** Games that incorporate blockchain-based assets. Ownership extends beyond the platform. This model attempts to merge play and finance.

75) **Digital assets:** Assets that exist primarily in digital form and are secured cryptographically. They include cryptocurrencies, tokens, and NFTs. Digital assets redefine ownership.

76) **Digital identity:** The representation of individuals in digital systems. Digital identity mediates access, trust, and participation. Control over identity becomes a form of power.

77) **Digital ledgers:** Electronic records that track transactions or ownership. Blockchains are a specialized form of digital ledger. Ledgers structure economic memory.

78) **Digital literacy gradient:** The uneven distribution of technological understanding across populations. Higher literacy confers economic and political advantage. Gaps widen over time.

79) **Digital Markets Act (EU):** A European law aimed at limiting the power of large digital platforms. It imposes rules on gatekeepers. The act reflects concern over market concentration.

80) **Digital Services Act (EU):** A European regulatory framework governing online platforms and content moderation. It emphasizes transparency and accountability. Enforcement shapes platform behavior.

81) **Digital wallets:** A software-based system that stores cryptographic credentials used to hold, send, receive, and manage digital assets and digital identities. Rather than storing money itself, a digital wallet controls private keys that prove ownership and authorize transactions on blockchain networks or traditional payment systems.

82) **Economics of attention:** A framework analyzing how attention is produced, traded, and exploited. It explains why platforms optimize engagement. Attention becomes an economic input.

83) **Electronic Numerical Integrator and Computer (ENIAC):** One of the first general-purpose electronic computers. ENIAC demonstrated computational scale and speed. It marked the transition from mechanical to electronic computing.

84) **Embedded finance:** Financial services integrated directly into non-financial platforms. Payments, lending, and insurance become invisible features. This reshapes consumer relationships with money.

85) **Embodied cognition:** A theory that cognition is shaped by physical interaction with the world. In AI, it challenges purely abstract intelligence models. It highlights limits of disembodied systems.

86) **End-to-end encryption:** A security method where only communicating users can access messages. Intermediaries cannot read content. Encryption protects privacy but complicates law enforcement.

87) **Enigma machine:** A mechanical encryption device used during World War II. Its decryption influenced the war's outcome. Enigma represents early cryptographic arms races.

88) **ERC-20:** A token standard on Ethereum defining how fungible tokens operate. It enabled rapid token creation. ERC-20 standardized digital assets.

89) **ERC-721:** A token standard for non-fungible tokens on Ethereum. Each token is unique. ERC-721 made NFTs possible.

90) **Ethereum (ETH):** A blockchain platform designed for programmable contracts. Ethereum expanded blockchain beyond currency. It underpins much of DeFi and NFTs.

91) **Ethereum (ETH) Layer 1:** The main Ethereum blockchain where transactions and smart contracts execute. It prioritizes security. Congestion has driven scaling solutions.

92) **Ethereum (ETH) Layer 2:** Secondary systems built to scale Ethereum. They reduce fees and increase throughput. Complexity increases trust assumptions.

93) **EU AI Act:** A comprehensive European law regulating AI systems by risk category. It imposes obligations on developers and deployers. The act sets global regulatory precedent.

94) **Executive Order 14110:** A U.S. executive order addressing AI safety, security, and governance. It directs agencies to assess risks and standards. The order reflects growing state concern.

95) **Exchange traded fund (ETF):** An investment vehicle traded on public exchanges. ETFs provide exposure to assets without direct ownership. Crypto ETFs bridge traditional and digital finance.

96) **Fiber optic cables:** Cables that transmit data as pulses of light through glass or plastic fibers. Fiber optics allow extremely high data speeds over long distances. They form the physical backbone of the modern internet.

97) **Fintech:** Financial technology refers to digital tools that modernize payments, lending, investing, and banking. Fintech replaces or augments traditional financial institutions with software-driven systems. It shifts finance toward platforms and APIs.

98) **Fifth Generation Warfare:** A form of conflict emphasizing information, influence, cyber operations, and economic pressure rather than conventional force. Digital platforms, social media, and networks become battlegrounds. Warfare extends into civilian life.

99) **Fungible Digital Assets:** Digital assets that are interchangeable and identical in value, such as cryptocurrencies or standardized tokens. Fungibility allows assets to function as money or commodities. This contrasts with unique assets like NFTs.

100) **Gas fees:** Transaction fees paid to blockchain networks to process and validate activity. Gas fees compensate validators or miners for computation and security. High fees reveal congestion and scalability limits.

101) **Genesis Block:** The first block in a blockchain. It establishes the initial state of the ledger. The Genesis Block has symbolic and technical significance.

102) **Generative Adversarial Networks (GANs):** A class of AI models where two systems compete: one generates data, the other evaluates it. GANs are used to create realistic images, audio, and video. They power much synthetic media.

103) **Generative AI:** AI systems that create new content rather than classify existing data. These systems generate text, images, code, and audio. Generative AI reshapes creativity and labor.

104) **GENIUS Act:** A proposed U.S. law focused on regulating stablecoins and digital assets. It reflects growing concern over monetary sovereignty and consumer protection. The act signals formal integration of crypto into policy debates.

105) **Governance tokens:** Tokens that grant holders voting rights within decentralized protocols. Governance tokens distribute decision-making power across participants. In practice, influence often concentrates among large holders.

106) **HTML (HyperText Markup Language):** The foundational language used to structure web pages. HTML defines how text, images, and links are displayed. It enabled the early expansion of the World Wide Web.

107) **Hyperaccumulative wealth dynamic:** A pattern where digital systems allow capital to compound faster than labor or regulation can respond. Scale and automation amplify returns for early or dominant actors. This dynamic drives inequality.

108) **Integrated circuit:** A compact electronic circuit containing many components on a single chip. Integrated circuits made modern computing possible by reducing size and cost. They enabled personal and mobile devices.

109) **Interoperability:** The ability of systems to work together across platforms or networks. Interoperability reduces lock-in and fragmentation. It is a key challenge in digital and financial systems.

110) **Large language models (LLMs):** AI models trained on vast amounts of text to generate and understand language. LLMs power chatbots, translation, and content creation. Their outputs reflect both capability and bias.

111) **Layer 1 blockchain:** The base protocol of a blockchain network. Layer 1 handles core security and consensus. Scalability constraints often emerge here.

112) Layer 2 blockchain: Secondary systems built on top of Layer 1 blockchains. They improve speed and cost efficiency. Layer 2 solutions trade simplicity for scalability.

113) Legacy drag: The resistance created by outdated systems, institutions, or processes. Legacy drag slows adaptation to new technologies. It often favors incumbents over innovation.

114) Lending pools: Shared pools of digital assets used for decentralized lending. Participants supply liquidity and earn interest. Borrowers access funds through smart contracts.

115) Lending protocol: A blockchain-based system that automates borrowing and lending. Terms are enforced by code rather than institutions. Risk management is algorithmic.

116) Machine learning (ML): A subset of AI focused on learning patterns from data rather than following explicit rules. Machine learning improves with more data and feedback. It underlies most modern AI systems.

117) Marginal cost: The cost of producing one additional unit of a good or service. In digital systems, marginal cost often approaches zero. This disrupts traditional pricing models.

118) Memecoins: Cryptocurrencies driven primarily by social attention rather than utility. Memecoins demonstrate the power of narrative and speculation. Their value is highly volatile.

119) Microdomain mastery: Specialized expertise within narrow technical domains. Digital tools reward depth over breadth. Mastery becomes more valuable but less transferable.

120) Miners: Participants who validate blockchain transactions through computational work. Mining secures networks like Bitcoin. It consumes significant energy.

121) Minting: The creation of new digital assets on a blockchain. Minting defines supply and ownership. It is foundational to tokens and NFTs.

122) National AI Research Resource (NAIRR): A U.S. initiative to expand access to AI compute and data for researchers. NAIRR aims to democratize AI development. It reflects concern over concentration of resources.

123) **National digital identity systems:** State-issued digital identity frameworks for citizens. These systems centralize authentication and access. They raise questions about surveillance and control.

124) **Network effect:** A phenomenon where a system becomes more valuable as more people use it. Network effects favor large platforms. They reinforce market dominance.

125) **Networks:** Interconnected systems that transmit information, value, or influence. Digital networks compress distance and time. They reshape social and economic structures.

126) **Neural network:** A computational model inspired by biological neurons. Neural networks learn through weighted connections. They power image recognition and language models.

127) **Neo-bank:** A digital-only bank without physical branches. Neo-banks emphasize user experience and low costs. They rely on existing banking infrastructure.

128) **Nodes:** Individual participants that maintain and validate blockchain networks. Nodes enforce rules and ensure decentralization. Their distribution affects security.

129) **Non-fungible tokens (NFTs):** Unique digital assets recorded on a blockchain. NFTs represent ownership of specific items or rights. They redefine scarcity in digital spaces.

130) **Occupy Wall Street:** A protest movement that emerged in 2011 in response to financial inequality and corporate power. It popularized the idea of the "99 percent versus the 1 percent." The movement foreshadowed later debates about platform power, wealth concentration, and digital finance.

131) **On-chain credit:** Credit systems implemented directly on blockchains using smart contracts. Lending decisions rely on collateral and code rather than traditional credit scores. This model prioritizes automation over discretion.

132) **Open source communities:** Groups that develop software collaboratively and make code publicly available. Open source has driven much of the internet's infrastructure. It contrasts with proprietary control.

133) **Oracles:** Systems that provide blockchains with external data such as prices or events. Oracles connect on-chain logic to the real world. They are critical points of trust and vulnerability.

134) **Organizational inertia:** The tendency of institutions to resist change due to structure, culture, or incentives. Inertia slows technological adaptation. It often protects legacy systems.

135) **Patent portfolios:** Collections of patents held by companies or institutions. Patent portfolios can be defensive or strategic. In technology markets, they shape competition and control.

136) **Peer-to-peer lending (P2P lending):** A lending model where individuals lend directly to one another through platforms. P2P lending reduces traditional intermediaries. Risk is distributed among participants.

137) **Phishing schemes:** Fraudulent attempts to obtain sensitive information by impersonating trusted sources. Phishing exploits human behavior rather than technical flaws. It remains a leading cause of digital theft.

138) **Platform economy:** An economic system centered on digital platforms that mediate interactions between users, producers, and advertisers. Platforms benefit from network effects. They concentrate power and data.

139) **Platform sovereignty:** The ability of states or societies to control the platforms that shape their economies and information flows. Platform sovereignty challenges the dominance of global tech firms. It is increasingly tied to national policy.

140) **Privacy theater:** Practices that appear to protect privacy without meaningfully doing so. Privacy theater provides reassurance rather than real control. It often substitutes transparency for power.

141) **Private blockchain:** A blockchain controlled by a single organization or consortium. Private blockchains prioritize efficiency and access control. They sacrifice decentralization.

142) **Private keys:** Secret cryptographic keys that grant control over digital assets. Whoever controls the private key controls the asset. Loss of keys usually means permanent loss.

143) **Proof of Stake (PoS):** A blockchain consensus mechanism where validators secure the network by staking assets. PoS reduces energy consumption compared to mining. Influence correlates with stake size.

144) **Proof of Work (PoW):** A consensus mechanism where participants compete using computational power. PoW secures networks through energy expenditure. It prioritizes security at high cost.

145) **Proprietary software:** Software owned and controlled by a company rather than the public. Proprietary models limit access to source code. They prioritize control and monetization.

146) **Protocol tokens:** Tokens native to blockchain protocols used for incentives, governance, or fees. Protocol tokens align user behavior with network health. Their value depends on adoption.

147) **Public blockchain:** A blockchain open to anyone for participation and verification. Public blockchains emphasize transparency and decentralization. They trade speed for openness.

148) **Public key:** A cryptographic key used to receive encrypted data or assets. Public keys can be shared freely. They pair with private keys for security.

149) **Ransomware:** Malicious software that encrypts data and demands payment for release. Ransomware targets institutions and infrastructure. It exploits digital dependence.

150) **Real-time bidding:** Automated auctions that sell advertising space in milliseconds. Real-time bidding uses behavioral data to target users. It monetizes attention at scale.

151) **Regenerative Finance (ReFi):** Financial systems designed to promote environmental or social regeneration. ReFi applies blockchain tools to sustainability goals. It blends finance with values.

152) **Reinforcement learning:** A machine learning approach where systems learn through trial and error. Rewards guide behavior over time. Reinforcement learning excels in dynamic environments.

153) **Regulatory arbitrage:** The practice of exploiting differences between regulatory systems. Firms move activities to favorable jurisdictions. Digital assets intensify this behavior.

154) **Regulatory capacity gap:** The mismatch between regulatory complexity and institutional ability. Technology evolves faster than oversight. This gap creates enforcement blind spots.

155) **Regulatory overhang:** Uncertainty created by pending or unclear regulation. Regulatory overhang can delay investment and innovation. Markets price uncertainty heavily.

156) **Regulatory territoriality:** The application of national laws to digital systems that cross borders. Territorial regulation conflicts with global platforms. Enforcement becomes complex.

157) **Robo-advisor:** An automated investment service using algorithms rather than human advisors. Robo-advisors reduce costs and standardize strategies. They prioritize efficiency.

158) **Robo-portfolios:** Algorithmically managed investment portfolios. Robo-portfolios rebalance assets automatically. They reduce human judgment.

159) **Seed phrases:** Human-readable backups of cryptographic keys. Seed phrases allow recovery of digital wallets. Their security is critical.

160) **Self-supervised learning:** A machine learning method where systems generate their own training signals from data. It reduces reliance on labeled datasets. Self-supervision enables scale.

161) **Self-sovereign data wallets:** Systems that allow individuals to store and control their personal data. These wallets aim to reduce platform dependency. Adoption remains limited.

162) **Semiconductor:** A material with controllable electrical conductivity. Semiconductors are the foundation of modern electronics. Supply chains are geopolitically sensitive.

163) **Sharding / Layer 2 solutions:** Techniques used to scale blockchains by splitting or offloading work. Sharding divides data across networks. Layer 2 reduces congestion.

164) **Silicon:** A chemical element used in semiconductor manufacturing. Silicon enabled microelectronics. It gave rise to the modern computing era.

165) Silicon Valley: A geographic region synonymous with technology innovation. Silicon Valley shaped startup culture and venture capital. Its influence extends globally.

166) Smart contracts: Self-executing code that enforces agreements on blockchains. Smart contracts reduce the need for intermediaries. Bugs can have irreversible effects.

167) Socially driven tokens: Tokens whose value depends on community participation or identity. These tokens monetize social behavior. Sustainability is uncertain.

168) Solana (SOL) Layer 1: A high-speed blockchain emphasizing throughput. Solana prioritizes performance. It trades decentralization for efficiency.

169) Spending telemetry: Data collected from consumer transactions. Spending telemetry reveals behavioral patterns. It fuels credit and marketing models.

170) Stablecoins: Digital assets pegged to stable values such as fiat currency. Stablecoins aim to reduce volatility. They bridge crypto and traditional finance.

171) Stakers: Participants who lock assets to support Proof of Stake networks. Stakers earn rewards for validation. Risk includes slashing penalties.

172) Staking: The act of locking digital assets to secure a blockchain network. Staking aligns incentives with network health. Returns depend on participation.

173) Status quo bias: A preference for existing conditions over change. Status quo bias shapes public reaction to technology. It resists both progress and regression.

174) Surveillance capitalism: An economic system in which personal data is extracted, analyzed, and monetized to predict and influence behavior. Surveillance capitalism turns human experience into raw material. It concentrates power in data-rich platforms.

175) Supervised learning: A machine learning method where models are trained on labeled data. The system learns by comparing predictions to known outcomes. Supervised learning requires extensive human-curated datasets.

176) **Synthetic assets:** Digital assets that replicate the value of real-world assets through code and collateral. Synthetic assets allow exposure without ownership. They introduce counterparty and systemic risk.

177) **Synthetic media:** Digitally generated content that imitates real people, voices, or events. Synthetic media blurs the boundary between authentic and artificial information. Trust becomes harder to establish.

178) **TCP/IP (Transmission Control Protocol / Internet Protocol):** The foundational communication protocols of the internet. TCP/IP governs how data is packaged, transmitted, and received. It enabled global networking.

179) **Technocrats:** Individuals who exercise authority based on technical expertise rather than democratic mandate. Technocracy emphasizes efficiency and optimization. It raises questions about accountability.

180) **Technological displacement anxiety:** Fear or unease caused by rapid technological change. Displacement anxiety reflects uncertainty about jobs, identity, and social roles. It influences political and cultural responses.

181) **Technological disequilibrium:** A condition where technology advances faster than social or institutional adaptation. Disequilibrium creates instability. Governance lags behind capability.

182) **Technological meditation:** A reflective approach to technology that emphasizes restraint and intentionality. It counters constant acceleration. This concept argues for human-centered adoption.

183) **Techno-historical mirroring:** The tendency to interpret new technologies through historical analogies. Mirroring helps understanding but can mislead. It simplifies novelty.

184) **Techonomics:** The study of how technology reshapes economic structures, incentives, and power. Techonomics integrates finance, AI, data, and governance. It frames the technological age as a distinct economic era.

185) **Terrestrial fiber networks:** Land-based fiber optic systems connecting cities and data centers. These networks carry most global internet traffic. Physical geography still matters.

186) **The FTX scandal:** The collapse of a major cryptocurrency exchange due to fraud and mismanagement. The scandal exposed governance failures. It accelerated regulatory scrutiny.

187) **The Transformer architecture ("Attention Is All You Need"):** A neural network architecture optimized for processing sequences such as language. Transformers rely on attention mechanisms. They power modern language models.

188) **Tokenization:** The process of converting rights or assets into digital tokens. Tokenization increases liquidity and programmability. It alters ownership models.

189) **Tokenized assets:** Assets represented digitally on blockchains. Tokenized assets can be traded globally. They challenge traditional custody systems.

190) **Tokens:** Digital units used for value transfer, access, or governance. Tokens vary widely in function. Their design shapes incentives.

191) **Transistor:** A semiconductor device used to amplify or switch electronic signals. Transistors replaced vacuum tubes. They enabled modern computing.

192) **Turing machine:** A theoretical model defining computation. The Turing machine established limits of computability. It remains foundational to computer science.

193) **Universal basic income (UBI):** A policy proposal that provides regular, unconditional payments to all individuals. UBI is often discussed as a response to automation and job displacement. It reframes welfare around universality rather than need.

194) **Unsupervised learning:** A machine learning approach where models identify patterns without labeled data. The system discovers structure on its own. This method is powerful but less interpretable.

195) **Validators:** Participants in Proof of Stake networks responsible for confirming transactions. Validators replace miners in PoS systems. Their incentives shape network security.

196) **Web 1.0:** The early phase of the internet characterized by static websites and limited interaction. Users primarily consumed information. Control rested with content publishers.

197) **Web 2.0:** An era defined by user-generated content, social media, and platforms. Interaction and participation became central. Data collection and network effects accelerated.

198) **Web 3.0:** A vision of the internet built on decentralized protocols and user ownership. Web 3.0 aims to reduce platform dominance. Its practical realization remains contested.

199) **World Wide Web (WWW):** A system of interlinked documents accessed via the internet. Created to share information globally, it became the dominant interface for digital life. The web transformed communication and commerce.

200) **Yield markets:** Financial systems focused on generating returns from capital deployment. In DeFi, yield markets are algorithmic and continuous. They expose leverage and risk.

201) **Yield strategies:** Methods used to maximize returns from assets. Digital systems automate yield strategies. Complexity often obscures risk.

202) **Zero trust architecture:** A cybersecurity model that assumes no implicit trust within a network. Every request must be verified. Zero trust reflects modern threat realities.

ABOUT THE AUTHOR

C. A. Chuba, known personally as **Corey Alan Chuba**, was born in Butler, Pennsylvania, and has spent his life rooted in the cultural and civic life of the Pittsburgh region. He served in the Pennsylvania Army National Guard and Army Reserves, earned a Master of Business Administration, and built his professional career in the corporate banking sector, while also pursuing entrepreneurial and nonprofit work.

A devoted husband and father, Chuba's interests include archery hunting, winemaking, gardening, and the study of classical literature, history, philosophy, theology, and geopolitics. He strives to order his life according to the Latin maxim *Omnia sunt sicut Deus vult* — "All is as God wills it."

Made in the USA
Monee, IL
10 November 2026

4f49ca27-690b-4a95-b71e-9446ab6cf845R01